AF841442

ODI Agricultural Occasional Paper 14

Managing Africa's Tropical Dry Forests:
a review of indigenous methods

Gill Shepherd

Overseas Development Institute

A CIP Publication data record may be obtained from the British Library.

ISBN 0-85003-139-7

© Overseas Development Institute 1992

Published by the Overseas Development Institute,
Regent's College, Inner Circle, Regent's Park,
London NW1 4NS

All rights reserved. No part of this publication may be reproduced by any means, nor transmitted, nor stored electronically without the written permission of the publisher.

Typeset in PostScript at the Overseas Development Institute
Printed by the Russell Press Ltd, Nottingham

Contents

Preface

A variety of sources were used for this work. For the ethnographic references, two major libraries had already been used in London for another Social Forestry literature review - that of the School of Oriental and African Studies, University of London, and that of the Royal Anthropological Institute at the Museum of Mankind, British Museum. Thirteen items from the material located for that study were relevant to this.

Secondly, species, forestry and agriculture references - and the documents to go with them - were sought through two major databases: those of the CAB (Commonwealth Agricultural Bureaux) and AGRIS, the FAO Forestry and Agriculture database. Librarians at the Science Library in London, the Oxford Forestry Institute library, and the Forestry library in Wageningen Netherlands were most helpful at this stage. Searches threw up large numbers of references of which most of the relevant ones were already in ODI. Nineteen new references were obtained from these databases.

Most of the documentation came from the Social Forestry library at ODI, built up through the Social Forestry Network, whose members receive network papers and newsletters in return for sending copies of their own writing, or project documents from the projects on which they work. The fact is that management references of the kind sought are often to be found in the grey literature which ODI collects, rather than in the formally published sources which enter bibliographic databases. Seventy-nine of the references summarised here were already in this library.

Three particular problems emerged from the searches; firstly, there are literally hundreds of species references for the chosen area which have nothing at all in them about either land tenure or indigenous management techniques; secondly, the ethnographic literature, while rich on tenurial arrangements, is poor on tree-species identification, and often on indigenous management as well; thirdly, it was unnerving, late in the survey period, to come upon Raison's work (1988) with its large francophone bibliography, and to discover how many of his references had not been thrown up by the supposedly multilingual databases already consulted. It is appended here to give it some circulation outside the francophone world, but the references within it could not by that stage be summarised or incorporated into the main work. This is a pity, but is unlikely to have materially affected the findings presented here.

Finally, a word about the production of the summaries. First drafts were produced by the author with the assistance of Joe Watts, Althea Ifeka, and Daniel Blais, two foresters and two social scientists. In order for the material summarised to be as comparable as possible, the main author created a set of headings under which each summary was composed, so that the absence of information on particular topics could be noted as well as its presence, and comparative information on particular topics relatively easily collated. The aim was to make each entry 100 - 350 words depending on interest, with a longer entry for especially important items. Topics were as follows:

1. **The physical conditions** of the area concerned, such as rainfall and type of tree-cover.
2. **Primary human economic activity**, such as herding, maize cultivation, etc.
3. **Tenure** - including competing concepts such as those of the state and local bodies, or local groups and individuals. Information on common property resources comes in here.
4. **The role of trees in the economic activities of the area**, including more minor cash-generating activities such as charcoal making or the gathering of non-timber forest products such as gum, fruit or nuts.
5. **The nature of management**: practices, control measures, rule-making, and rule enforcing.
6. **Adaptation to change** - to worsening ecological givens, to changing government strategies, to new competing uses of the area.
7. **Other, especially management proposals** for the future. (Or any other interesting material which does not fit into the above categories.)

The final selection and editing of summaries, and the analysis made from them, was the work of the main author. The book was prepared for publication by Moira Reddick and Peter Gee.

Part I

1. Introduction

This occasional paper presents the results of an extensive literature search and analysis of indigenous forest management practices in dryland Africa. The impetus for the task came from several directions.

Ever since the benchmark of the Eighth World Forestry Congress in 1978, a slow process of change in the thinking of forestry professionals has been going on. Where once their primary task was seen as the protection of forests and the maximisation of revenue for industry and the State, it has now become equally important to consider the dependence of rural people upon tree products and the effect upon them of forestry activities. Three factors lay behind the change. Firstly the oil price rises of the early 1970s made it clear that woodfuel was going to continue indefinitely to be the primary energy source of billions of rural people. Secondly it became apparent that the world's forests, which in earlier decades had been assumed to be an almost infinite resource, were rapidly disappearing, especially in drier areas. Thirdly, development thinking itself was increasingly concerned with the needs of the small producer.

Tree-planting projects with farmers and villagers were initiated in many parts of the world, largely in response to these concerns. Often referred to as social or community forestry, many such projects made the establishment of village woodlots their first goal. Results have been very variable, however, particularly in Africa, which most concerns us here. It has gradually become clearer that rural people want to plant trees mainly on private, permanently owned land; and only where land in general is so short that there is no bushland or forest nearby from which tree-products may be taken without the trouble of growing them. In Africa in particular, these prerequisites currently rule out the likelihood of successful villager tree-planting projects in many areas, particularly the drier, less densely populated ones. Woodland and bushland, though dwindling, still cover millions of hectares and will continue to be used by the populations who live near them for most purposes for the foreseeable future.

At the same time, plantations of exotic species have been less successful in dry areas than was expected. It is now clear (which it was not twenty years ago) that woodlots and plantations are unsuccessful in areas with less than 800mm of rainfall, except where trees can be irrigated (Lamprey 1986:125). 'Fast-growing' exotics grow scarcely faster under these conditions than the pre-existing vegetation and are outperformed by indigenous species in drought years. Managing woodland for fuelwood rather than establishing plantations becomes economic once plantation yields are below six cubic metres a hectare annually (Wormald 1984 6.3), a figure quickly reached if water is unavailable or plantations poorly managed.

As a result, the low esteem in which bushland used to be held is giving way to the recognition that previous estimates of natural woodland productivity may have been too low (Jackson 1983:12) and that, even at low increment levels, there are millions of hectares to be exploited. Most indigenous arid zone trees and shrubs continue to grow even when browsed, burned and lopped, and more protection leads to much increased productivity. However, good rains for Sahelian tree-establishment only occur every decade or two and woodlands tend to consist of large numbers of even-aged trees. If more young seedlings and saplings could be helped to survive until they were well-established, it has been estimated that such protection would be hundreds of times more effective in reafforesting drier areas than attempts to plant trees could ever be (Lamprey 1986:126).

In the colonial period in Africa the simple low-cost management of large tracts of existing woodland was the main form of government forestry practised. (The main activities were boundary demarcation and boundary protection). After a thirty year hiatus from the 1950s to the 1980s, during which the raising of plantations for industrial uses took precedence, the pendulum has again swung back and natural woodland management has become the object of renewed interest, dry savanna woodland being a particular priority.

However, much has changed since it was last tried. Africa's population has increased, while rainfall levels have decreased in the dry woodlands, intensifying competition for resources. Agriculture, for instance, has greatly expanded at the expense of pastoralism and woodland cover.

The greatest potential change of all lies in the evolving relationship between foresters and villagers, which increasingly rules out coercive solutions to the management of natural woodland. The difference this time is that social forestry has given donors and forestry professionals experience in collaborative activities with local people and a new confidence in their skills, whereas before the role of local people was simply to accept exclusion gracefully. Several projects, discussed here, have made the deliberate attempt to involve local people in the management of natural woodland adjacent to their homes.

A role for local people in forest management is also coming to be seen as just because it is slowly being realised that they already have complex knowledge in this area. In agriculture, the value of indigenous knowledge of natural resources has for some time begun to be recognised and given a place. Hardly surprisingly, in colonial times, official policy was almost bound to belittle the knowledge of local people - how else could a period of colonisation be justified? - but that was also a period in which it was assumed that western agricultural knowledge was likely to transform African agriculture in due course, and that therefore indigenous knowledge was old-fashioned knowledge. Only gradually, and in the more difficult and fragile environments, above all, did it become clear that indigenous knowledge had far more to teach than western knowledge. In these situations, a new respect for rural people and their solutions for difficult environments has been engendered, and has generated such research methods as farming systems research. In forestry, recognition of the extent of indigenous knowledge has come more through social forestry projects than any other route.

Only perhaps in social anthropology does a recognition of the complexity of indigenous knowledge go back even further in time - and for that reason some relatively early ethnographies are featured here.

The rationale for the work

This search and analysis was undertaken, then, mainly to encourage the forestry profession to involve those who live in or near to Africa's tropical dry forests more closely in their management, and to be more aware of their pre-existing knowledge and practices. It was assumed that better forest management plans might then be drawn up. Anglophone and francophone forestry, biology and ethnography materials were searched along with the grey project literature which is collected in ODI on participatory forestry. In all, stand alone summaries (essential for hard-to-obtain items) were made of 111 documents.

Redefining forest management

Management involves a series of mechanisms, put into practice by rural people who in many cases are coordinating their actions with others, at the command of some (ideally) local authority they regard as legitimate. In many cases, management is conducted almost entirely by people with rules in their heads and without coercion. In other cases, it is as important to know **why** people are obeying rules, and the sanctions involved, as it is to know what the rules are.

To the author's surprise, it became clear, as the review[1] went along, that we allow ourselves to fall into a trap if we talk about the management of woodland as distinct from the management of trees on farmland.

Firstly, in well over a third of the case examples (41 out of 111) specific tracts of land alternate over time between woodland and farmland, as swidden-fallowing[2] cycles are operated during which a woodland regrowth phase is allowed to replenish the fertility of the soil after a farming phase.

Secondly, we have to distinguish the woodland management practised by the herder from that practised by the farmer. For the herder, the woodland is a primary and immediate economic asset, as important as the animals herded themselves, since they must browse rather than graze for most of each year. For the mobile herder, the forest **is** his farm. For the crop-growing farmer, on the other hand, the forest is somewhat more in the background, on a day to day basis; it provides resources complementary to primary economic activity rather than resources primary in their own right, and management may well be related to the nature of forest as farmland-in waiting, and to the knowledge that farmland will return to fallow in due course.

Thirdly, many management practices, it turns out, form a seamless continuum from management in the forest through to management on the farm. Both forms of land are effectively held by the same corporate group, with individual rights the predominant, but not exclusive, mode when land is being farmed, and group (usually descent group) rights

---

1. Many of the articles used here are summarized in the ethnographic present, in line with the article itself. It is vital to check the date of the item. In this analysis, the ethnographic present is similarly used for the sections on management practices which were still extant at the time they were described. However, as the succeeding sections on change and the future make clear, we should not assume in too sanguine a way that all management practices are still functioning.

2. The hybrid term swidden-fallow has been used for two reasons. Firstly, it links two terms which, rather arbitrarily, are normally used respectively for non-African and African situations. Secondly, it stresses (in a way terms such as 'shifting cultivation' and 'slash-and-burn agriculture' do not), the fact that farmers do not abandon cleared land when they leave it to restore its fertility and plant crops on new land. They obtain other products from it, may retain continuing control over trees growing up on it, and will return to it when their own particular cycle is complete.

reasserting themselves when land is under bush fallow. It seemed a pity not to attempt to capture this process in the summaries chosen here, and so some material about the management of individual trees on farms is included along with that on the management of trees in forests.

Finally, it is now standard practice to deplore the inability of Third World Governments, and colonial regimes before them, to distinguish between common property resources and open access land. We would be unwise to make the same conceptual mistake by lifting the woodland managed by local people from the matrix in which it exists - the economic activities of farming or herding - and by treating it as if it were a forest reserve. By and large, the tenure and authority regimes which once governed the successful use of forest, have also been those that govern the use of farmland and all other local resources. It is the divorced management of the two respective resources - by different ministries, by ordinary people or officials, by local and non-local authorities - which has led to many present-day problems.

The starting point for this paper then, is that ownership and management go together, and that in many cases farmland or rangeland, and woodland, cannot be separated. Only by understanding tenure fully, will we understand the conditions for successful management.

It is important, too, to be aware that 'indigenous management' takes at least three broadly distinct forms, within which are much larger numbers of actual practices. As with the forest management practised by professional foresters, it is important to ask, 'management for what?' in the case of these three forms.

Herders manage the largest tracts of forest, patrolling a wide domain with their animals and attempting to preserve and own perennial dry season grazing if possible. They try to keep a wide range of species alive, and are interested in a range of fodders which are available at different times of year, suit different kinds of stock, and vary in their nutritional content.

Sedentary farmers manage woodland most intensively nearest their home bases, and with decreasing intensity, in a series of concentric circles as one moves outwards from this base. Such 'graded' management needs to be understood since it is rather different from a forest reserve management with equal intensity throughout.

Trees on farms are the most intensively managed trees of all. The planning and management of the individual tree seen in this context, is rarely matched by any such actions in communally owned woodland.

2. Ownership of woodland

Management of natural woodland is practised by those to whom it belongs, and, as has been seen so clearly in the case of tree-planting, no serious investment of time and effort will be made unless the resource is owned. Thus the mechanism for ownership must be our initial focus.

The herding lineage

In the case of many of Africa's herding groups, the genealogy of the lineage is the charter for access to land. All the male descendants of one ancestor share one large area, but subgroups of more closely related kin intensively share smaller portions for certain purposes. Thus any one piece of land is shared by small numbers of owners with strong claims and large numbers of owners with weak claims to it. Cases summarised here such as the Libyan Bedouin [11], the Baggara of the Sudan [12] and the Turkana of northern Kenya [6; 7; 8] demonstrate the flexibility of this system. The Libyan and Turkana cases in particular show how tenure is related to predictability of rainfall or access to perennial water: access narrows when a perennial asset is at stake, and broadens where low and erratic rainfall makes it a gamble where the best grazing will be from year to year.

In addition, of course, agreements have to be forged about the boundaries between adjacent tribes or clans[3]. These were created in the past by political and military means and have always fluctuated to a degree [109]. In Somalia's Bay Region, elders explicitly linked the right of access to grazing land to the duty to fight for it when need arose [94].

This 'nesting' model for lineage territory, with land belonging to tribe/lineage/lineage-section/and immediate kin-group but with the focus dependent upon context, is at its most flexible when it applies to fully mobile pastoralists, though as an ideology it is also found in modified form among semi-sedentary groups (e.g. northern Kenyan Samburu on group ranches - [76]) and indeed among agriculturalists (e.g. the matrilineal Aouan of Côte d'Ivoire [20]; the gum farmers of Kordofan, Sudan [90]).

---

3. These terms tend to be used vaguely. In fact, a tribe is a large category of people who speak the same language and assert, but cannot trace, kinship with one another. Clans are sub-categories of tribes with an area focus, but are still large and vague. There are only five clans in the whole of Somalia, for instance. It is at the lineage level (the clan sub-section) that members can reckon their relationship to one another by reference to a common ancestor.

In the case of mobile pastoralists, the actual extent of the range owned by the lineage will vary over time, and individuals with their herds will find themselves in different parts of the current range at different times of the year, or in different years. There is no necessary relationship between individuals and specific pieces of land as there is in the case of farmers: membership of the lineage is enough, and flexibility after that is considerable.

Sedentary kinship groups

For sedentary farmers, while a knowledge of genealogical links is important, the 'moral community' which holds land is defined by both descent **and** residence, rather than descent alone. Very often this is expressed as the chief holding the land on behalf of the collective, and the lineage members who live in his area drawing usufruct rights through him. This is the case in the current collection of summaries, for the Unyamwezi of Tanzania [1]; for areas of southern Ghana [65; 71]; for Kakamega district in western Kenya [27]; and for the Peul of the inner Niger delta in Mali [98].

Alternatively, we are told that the clan is the descent-and-residence group and often this means that no single chief allocates rights, but that he and groups of elders do. This seems to be the case for the Gusii in Western Kenya [5]; for the Bay region Somali [94]; in eastern Botswana [57; 88; 96] and in various parts of Niger [28].

The village itself is at times the focus of descent-and-residence rights, with the village headman (or sheikh) making land allocation decisions over which the clan chief - a superior in descent terms but a non-resident - has no control. We find this among the Nyakyusa [44]; and the Sukuma [64; 95] of Tanzania; and in the northern Sudan [46; 48].

Finally, there is mention of village management of resources without it being clear if and how descent and residence rules work. The Dogon of Mali operate a management system run by representatives of village quarters [104]; and in Okwe, southern Nigeria, adjacent villages manage aspects of their swidden-fallows as a collectivity [39].

The household head and the household

Just as the lineage fits within the tribe, and the village is often as a subset of the lineage defined by residence, so the household is often seen merely as the lineage writ small. This is clear from the fact that land would revert to the next collectivity upon the owner's death, rather than to his children. There are examples here from the Sudan [46], Ghana [65] and among the matrilineal Luguru of Tanzania [13].

In some systems, the household head used in the past to have to accept that co-villagers had some rights to self-seeded trees on his land, but, like the larger groups of which he was a part also, he could limit the rights of outsiders and control the access of different categories of insiders [80].

However, the household head's position is gaining in importance all the time, as other levels of the kinship system cease to have political meaning and as land registration becomes the norm [34; 35; 73; 109]. In the future, user groups for particular natural resource assets are more likely to be aggregated upwards from the household, rather than a subsection of a larger collectivity.

The right to manage: the types of actions undertaken by rightholders

For chiefs and lineage elders of herding lineages, management actions most commonly undertaken are likely to be the exclusion of outsiders, [93; 94], adjudication between insiders, and the promulgation of new rules. Outsiders need to ask permission before bringing animals to graze within the lineage area [7]; or should stay right away if, as do charcoal burners, they have competing designs on the trees [96]. Samburu elders [76] decided to allow poor local men temporary permits to make charcoal, but banned charcoal making by outsiders, for instance. Turkana elders issued new instructions about the lopping of *Acacia tortilis* (in consultation with project staff) so that only side branches were taken, and the leader shoot stood a chance to grow rapidly above goat-browsing height [60].

Chiefs, elders and village leaders in villages seem to have attempted a wide variety of management initiatives. Many areas with strong chiefship or even kingship institutions have attempted to preserve the most valuable tree species in the area and linked them to the very office they held. In some cases the attempt was linked to an altruistic attempt to encourage farmers to keep alive the species they most needed on their swidden-fallows, in others a personal perk seems to have been the driving force behind the edict. Examples of the former include the protection of *Faidherbia albida* [25; 43] in Niger and in the Sahel in general [99]; and the protection of *Acacia senegal* for gum arabic in the Sudan [45]. An example of rulers seeking to extract personal benefit from control of species associated with the chiefship occurs in northern Ghana, where the species in question are *Parkia clappertonia* and *Butyrospermum parkii*, both species producing a commercially valuable product [70].

In south-eastern Botswana, village chiefs would ban the felling of village amenity trees (for shade, and lavatory shelter for those too elderly to go out to the bush [86;88]), and arranged elaborate zoning for different categories of fuelwood collector [96]. Around Mount Kenya (and widely elsewhere in Africa) sacred groves were used as the meeting places or burial grounds of chiefs or senior age-sets, and to the extent that these groves symbolised the patrilineage, women were banned from ever entering them. In the Côte d'Ivoire [20] elaborate arrangements for protecting and resting the forest were originally in force. Punishments for forest misuse, too, are often graded, and villagers are likely to be made to pay lower fines for local crimes than are outsiders [104].

At individual household level, the lineal household head controls access to the trees on his land: makes rules for his wife and neighbours (both categories of insider-outsiders) as to the species that may be cut (e.g. [35] for the Kamba in Kenya); or the thickness of branches that may be cut (e.g. [80]; for a Wolof village in Senegal) or the volume that may be cut (e.g. [42] for the Kenyan Maasai); or indeed whether any cutting will be allowed at all [73], just as lineage heads do for would-be immigrants and clients.

The rights of secondary users

What we see across all these cases is a system perhaps most readily described as a series of background and foreground rights. In the background are broad rights in forest and farmland with the potential for crisper definition, while in the foreground are firm rights, narrowed down to a smaller number of right-holders, which have been won by deliberate action. Clan and lineage mechanisms manage large areas lightly [3; 5], while smaller groups consolidate their rights by specific action.

At the same time, secondary rights remain very important. Among pastoralists, for instance, one of the risk-spreading actions has always been to marry daughters to remote lineage sections or to a man from another lineage entirely, in order to gain secondary, in-law rights to other areas. Limited rights can be a great deal better than no rights at all, as is clear from the Burkina example [68] in which herders are allowed to graze on farmers' fields after harvest, but not (as the farmers themselves do) lop browse from farm trees for their own animals.

In Senegal, all co-fallowing villagers may take firewood from the one another's land in the dry season, only wives may collect from their own husband's fields in the wet, and only the farmer himself can cut larger branches, whatever the season.

3. The creation of tenure through labour

Several of the summaries make it clear that it is the investment of labour which creates ownership. In the case of land, this means, for the agriculturalist, being the first to clear and plant land once under forest or woodland [10; 14; 37; 65]. In all cases until present-day land registration procedures, such cultivated land reverted if it was abandoned, not to the wild but to the group to which the clearer belonged [13]: so the individual had created rights for others as well as himself. The creation of pastoralist tenure, implicit in the Turkana examples [6; 7; 8], takes place in the dry season, and is manifest in the successful defence of the key dry season assets of grazing-browse and water. Here labour investment may mean simply defence, though in some areas the annual restoration of wells or clay-lined reservoirs [94] is a key tenure-defining piece of labour.

But labour creates ownership in all kinds of ways. Banging pegs into a tree suitable for hanging honey barrels in, among the Tanzanian Sukuma, ensures exclusive access to the tree [64]. Cut and stacked piles of firewood in the Samburu forests in Northern Kenya will not be touched by other women [76].

It is clear that tenure has to be worked at in some of these cases and that, as a general rule, individual tenure comes from the most, and the most constant, labour investment. In the Senegal case [80], land was lost to the owner if it remained uncultivated for more than ten years, for instance, and similar potential loss of tenure of dry season grazing is reported from the Turkana [7]. In the case of the Cyrenaican Bedouin, investment in water-points greatly strengthened tenure [11].

Tree-planting, because it is more work than tree-use, also creates tenure. Planters of valuable or exotic trees [65] strengthen their rights to the land on which the trees are planted in most cases - mainly because they are replacing seasonal with perennial crops.

Note, however, that the investment of labour must come from its owner's own free-will. Neither slaves [101] nor tenants [79] can easily create ownership for themselves by their labour, since it has already been made clear in their case that their labour belongs to another within the terms of their status, or their contract with the land-owner.

4. Indigenous woodland management methods

Management in the sense in which it is used here means some individual or group activity which organises the utilisation of tree resources in such a way that the resource is more equitably shared, more likely to remain into the future (more sustainably used) or will grow better than if the management practice was not taking place.

Classical planned natural woodland management, of the kind practised in European forests in the past, is somewhat different. Firstly, as indicated above, forest and farmland were early separated, and grazing in forests has by no means been universal.

Secondly, the drafting of management plans for rotations of particular lengths and specified products is naturally unknown in rural Africa. Though it will be seen from systems described here that planned actions which encourage some species and eliminate others, or which encourage trees to produce different end products - thin poles, thick timber, for instance - are by no means uncommon. However, many of them take place in the context of farming, rather than dedicated forest.

What they **are** doing is devising rules for sparing certain species, or certain size classes, or simply saying who may and who may not use certain tracts of woodland. They may mean management of the individual tree, but more often mean management of the space which the trees occupy, along with the grass and water there too.

Here, all and any planned and deliberate activity which enhances the quantity and quality of woodland or makes its use more sustainable, is defined as management.

Long and short swidden-fallow systems

Swidden-fallowing has been the most important method of woodland management in Africa for many hundreds of years. The selection of summaries here, while giving some flavour of the variety of patterns and types of fallow to be found in the continent[4], is only a minute fraction of a vast literature on the subject.

Order can be brought to the array partly by focusing on the types of swidden-fallow associated with particular tree-species, and partly by attempting to identify the management procedures which governed the progression of the fallows.

One of the most useful documents identified during the literature search is that by Raison, 1988 [84]. It deals with the Sahelian and Sudanian zones of Africa, (among others) where the inhabitants practise both agriculture and livestock raising, and where two particular

---

4. The summaries which mention fallowing are numbers: 2; 4; 7; 10; 14; 15; 20; 21; 25; 28; 29; 30; 31; 38; 44; 46; 47; 49; 50; 52; 55; 62; 64; 68; 69; 72; 73; 75; 76; 77; 83; 86; 87; 90; 92; 93; 94; 102; 105; 106; 108; 112.

swidden-fallowing systems predominate. In these zones, settlements create a variety of 'parklands' placed in concentric rings around the village, which are constantly evolving. Leaving aside the permanently cultivated village home-gardens, we find two zones before the unadapted bush is reached:

- the zone of permanent fields set with *Faidherbia albida*;
- the zone of fields cleared in the bush, and of long swidden-fallows, where *Butyrospermum parkii* (*Vitellaria paradoxa* or karité), *Parkia biglobosa* (néré) and *Ficus platyphylla* are found.

The descriptions given here of the two are summaries of Raison's insights.

Butyrospermum parkii parkland - bush swidden-fallow (long fallow). (References in the collection of summaries: 14; 28; 29; 30; 31; 38; 49; 54; 68; 70; 75; 83; 84). This formation is found throughout the Sudanian zone apart from Senegal, and is associated with long swidden-fallows. In the untouched wooded savannah, with initially perhaps 1000 trees per ha (diameter range 5-60cm) all but 100 or so are burned so that they die. *B. parkii* are the main species preserved. Over the next four years, much of the dead wood is used up for fuelwood, as the area is cultivated, and then the whole patch is rested for 20 years and the farmer moves on outwards from his village. If one area is re-used by one village for 2-3 successive cycles, i.e. 40-60 years, with the *B. parkii* protected all that time, the shade after sixty years will be so great that agriculture will suffer. So the whole village will move, using the earlier area as a fruit-gathering area, and the process will start again. After 2-3 moves - 150-200 years - the *B. parkii* will be very elderly, and it will be time to start the whole process of major land clearance again, preserving younger trees.

In the documents identified for this survey, references to the manner of fallow management at village or group level are scanty. It is implied that, in Niger, land management systems were organised by the village chief, elders and household heads [28], however. What is impressive in the *Butyrospermum parkii* cycle is the lengths of time involved, the coordinated nature of village planting and protection, and the extent to which the landscape in these regions is man-made.

Faidherbia albida parkland - savannah swidden-fallow (short fallow)[5] (References in these summaries; 2; 7; 10; 14; 25; 29; 30; 38; 43; 49; 51; 54; 61; 68; 74; 75; 80; 83; 84; 87; 89; 92; 99; 102; 105). This is the most interesting and developed of the types of parkland created by Sahelian farmers. The tree, because it is in leaf in the dry season, and leafless in the rains, permits permanent cultivation under it, and also, at a density of 10-30 per ha, fertilises up to 50% of the area. Crop rotation is practised under the trees.

However, it is clear that *F. albida* alone cannot restore soil fertility - animal dung is needed too. A hectare of *F. albida* parkland with twenty trees on it supports six cattle in the dry season - enough to keep the area fertile; but in the rainy season the tree is leafless, there are no agricultural residues available for fodder, and the bushland is then an essential complement. If it is nearby, livestock can browse there and be tethered on currently exploited fields at night to deposit their manure. If there is no bush nearby, more labour-intensive solutions must be adopted, such as sending animals north for part of the year, or growing forage for them. But more recently, with the departure of the young as labour migrants, animal herding has become too labour-intensive an activity. The result is often that *F. albida* parkland fails and reverts to less intensively managed *Butyrospermum parkii* parkland.

There are equally few clues to the style of management of these parklands, though several writers have noted the symbiosis between the tree and cattle - *F. albida* making it possible to keep stock on the land in the dry season resulting in the additional fertility of the field from manure [61; 74].

Acacia senegal gum-gardens. (References in the summaries: 28; 45; 46; 83; 84; 87; 89; 90; 99; 109). In the case of gum arabic, we lack Raison's clear descriptions, and our understanding of the swidden-fallowing system is thus patchier. It would seem that each household was allocated (by the village sheikh in the Sudan) three or four plots of land, each of which was farmed for 3-6 years on a rotational basis. *Acacia senegal* naturally regenerated in cleared plots and was then protected as it grew up. Very occasionally, *A. senegal* seeds were sown. As the trees came to

---

5. Hans Ruthenberg, (*Farming Systems in the Tropics*, Third Edition, 1980, Clarendon Press, Oxford) would call the *Faidherbia albida* parklands, with their heavy dependence for proper functioning on the systematic incorporation of cattle and their dung into the system, unregulated ley systems.

maturity on the fallows they were tapped for gum by those whose fallows they were [45,46]. Finally, when the trees were old, they were felled and sold as charcoal. It is not clear from descriptions whether the circular swidden-fallowing pattern around settlements was the norm, or whether farmers' fallows and cultivation were arbitrarily placed. Centrally allocated gum plots would suggest the former. The situation is collapsing - partly because population densities are now too high to allow such large amounts of land per household.

Private fallow systems?

Much fallowing in semi-arid Africa would seem to have involved village-wide agreements about how the surrounding land was to be used, not least so that the whole settlement could ultimately decamp together. A few documents stress the private nature of particular swidden-fallowing systems as well. The example provided here is the Bougagé strip rotation system in Niger [89]. Here the household controlled 17 strips of which at any time, 12 are being fallowed, 4 are in their first, second, third or fourth year of cultivation, and one is a house-site and cattle-pen. While it is possible that some groups practised long-swidden-fallowing in a more 'private' way than others, it is much more likely that different writers have focused on different things. While fallowed plots were most likely to be returned to by the household who had cleared them in the first place, the pattern of each household's fallows would have to fit with village-wide patterns as well.

The conservative use of bushland resources

The conservative use of resources sounds like an activity any individual would undertake without even thinking about it. However, when we find such attempts in large areas of woodland, they not only let us know that we are in the presence of a common property resource, but also that some kind of authority in the background is passing on the rules of good resource use and checking to make sure that they are obeyed. Among the instances noted in these summaries we find the following:

First, and most basically, fuelwood is a lowly end for wood and fuelwood ought to be dead wood. Live trees should be cut only for more important purposes, or for land clearance for agriculture [36; 45]. Of course, the temptation to cut live fuelwood will eventually come as resources become remote [70].

Secondly, in pastoral areas, thorny temporary fencing is essential at night to guard animals from predators. For this purpose, too, low-value, bushy, positively over-abundant species should be cut such as *Acacia reficiens* and *A. brevispica* according to the Turkana [6; 36]. However, even these species are too valuable for fuelwood unless they are dead.

Thirdly, important fodder trees must not be diminished by being used for less vital needs [35;76]. A wide range of fodder trees are needed to cope with the forage needs of several different species and ages of stock, to provide browse in the rainy and in the dry season, and for the production (variously) of meat or milk [6].

Fourthly, special-purpose trees such as *Salvadora persica*, widely grown for toothbrushes, should not be wasted on ends for which a wide-range of species would do [100].

Selective maintenance, and the promotion of particular species

Most areas in semi-arid Africa have begun by enjoying a surplus of trees, so that the people who lived there had the luxury of taking out unwanted species and concentrating on the preservation and enhancement of others [109]. The result has been a woodland composition strongly affected by human selection in the past [54]. We have already seen this process at work in the section on fallowing (4.1). Other species which have clearly been promoted in this way deserve a mention as well.

Summaries mentioning the fruit *Parkia biglobosa* (28; 29; 30; 31; 38; 49; 54; 68; 75; 84; 89; 102) show that it was enthusiastically preserved in the same swidden-fallowing systems as *Butyrospermum parkii*.

Adansonia digitata (14; 28; 29; 30; 38; 49; 54; 74; 75; 80; 84; 89; 92; 99; 102) is frequently promoted, planted and saved above all around villages, where it is valued for: its edible spinach-like leaves; its ascorbic fruit; the rope which can be made from its bark; and - in the Sudan - the water storage potential of its hollowed out trunk. It is also a good honey-barrel tree because it is not spiny to climb. The clumps of this species frequently found standing alone in the savanna are archaeological memorials to long abandoned villages [92].

Among many other valued trees *Borassus aethiopum* (39; 74; 75; 89; 91;92) is valued for basket-making; and *Balanites aegyptiaca* (6; 10; 38; 45; 68; 73; 75; 80; 83; 84; 89) is important for fodder, dry season fruit, as a leafy vegetable; and for construction, tools and utensils. It is occasionally even regarded as sacred [84].

Valued trees are cut and pollarded by pastoralists in such a way that they will live to be cut again [6;9]. Species that tolerate (even thrive on) lopping are encouraged such as *Balanites aegyptiaca* among the Pokot in Kenya [73], and *Combretum nigricans* in Guesselbodi Niger [47] and many more.

There is much evidence of the attraction of indigenous fruit trees in these summaries, from those preserved in Zimbabwe [23] to the Sahelian-Sudanian species such as *Tamarindus indica* and *Sclerocarya birrea* as well as the others mentioned here.

From Senegal to the Ethiopian Rift, and extending southwards as well, we find similar tree species preserved among crops, and similar species left and planted to enrich the woodlands [50;75].

As usual, there is far more evidence in the literature of **what** has been preserved, than there is of **how** it was preserved. As far as group management is concerned, we have to construct a picture from the few good accounts we have. Brokensha and Castro [22] have an excellent description of an island of forest managed by the elders in Mwea, Mount Kenya on behalf of rightholders who had to get permission to fell timber trees in it. We hear of effective village management of nearby woodland in the northern Sudan [48] and in northern Ghana [70], and of the protection of trees on farmland, even in the dry season, in Burkina Faso [68]. There is more data on individual management of trees on farmland or fallows, but we know that such action is organisationally easier [38; 73; 92].

Reservation and sacred groves

Considering this volume's thesis that trees are not set aside in indigenous management systems, but are lived among and used to the point where it is hard to say where woodland ends and farmland begins, it is surprising that reservation exists at all. Yet it does.

Many accounts make it very difficult to say whether trees are actually being reserved for soil conservation or watershed management reasons and the ban is merely wrapped in religious language to give it force, or whether the preservation of the trees is nothing to do with 'rationality' of this kind. It is certainly worth noting that in all Bantu African languages there is a unique noun prefix for all trees, gods and spirits and that many tree-species are thought to house spirits. Summaries [52] and [63] makes it clear, too, that such veneration goes beyond Bantu boundaries.

Some sacred groves acquire their importance from the fact that ancestors' graves are clustered in them, and economically or socially important species are clustered there as tombstones. It is customary to put such clusters of graves on a hilltop or ridge in some cultures - and so the trees may have an inadvertent conservation effect as well. Among the Kikuyu, such sites were commonly adorned with *Ficus natalensis* trees

[22; 58][6]. Among the Lozi of Malawi a small grove was planted at a king's grave with seedlings transplanted from the bush [41].

Some sacred groves (with their graves?) clearly symbolise the patrilineage, and women may not enter it [68], just as some especially huge trees with their many large and smaller branches may do [73].

At the practical rather than the sacred level, there are a few instances here of reservation for patently economic reasons. The grazing and browse reserves of the Pokot and Turkana are one such example from primarily pastoral areas [6;7;60], while the hilltop grazing and woodfuel reserves of the Sukuma in Tanzania are a more agricultural example [95].

Finally, in the past, some groups planted and guarded particular tree species as a reserve against famine [54; 84].

Management by time and season

In both agricultural and pastoral areas, it is a common-place to say that there are great differences between the dry season and the rainy season.

From the point of view of tree-management, trees in swidden-fallows and on fields are protected while there is a growing crop, but lie open, as does the land itself, to all comers in the dry season. For pastoralists, there is a plethora of browse in the rainy season, but the retention of browse sources for the dry season is of paramount importance.

The agricultural point is made by the description, for Senegal [80], of women's exclusive access to minor forest produce (MFP) from their husbands' fields in the rainy season and their competitive access, with all other villagers, in the dry season. In Botswana, 'to prevent hail' (so as not to damage the crops?) the chiefs used to put a ban on the cutting of certain tree-species (we are not told which) from land-preparation to harvest [86;88]. When the harvest was in, and the slack, house-repair season about to begin, the chief would 'open' the tree-cutting season again by making the first cut and insisting on repair to the *kgotla*, the village meeting place, before private repairs were begun. We hear too of open and closed cutting seasons in other areas, e.g. Marsabit in North East Kenya [100] and the management of common property resource

---

6. The Kikuyu appear to be named after this tree species - the Mkuyu - and their name can be translated as 'the little *Ficus natalensis* trees'. One wonders whether Kikuyu elders used to wear bark-cloth as did the Azande in the southern Sudan [37]. There, only men made and wore the cloth, and only they propagated the species, as a result.

(CPR) grass destined for thatching in Mwanza, Tanzania [95] by the setting of a cutting date.

For pastoralists, the earmarking or enclosure of dry season fodder is the key response to seasonal vulnerability e.g [6;7;8;60] for the Turkana; and [12] for the Baggara of the Sudan. Many pastoralists also leave certain categories of fodder till last - that near their home wells where they will return to wait out the last of the dry season, or that high up in trees where it is inaccessible to animals, but which can be lopped for them as a fall-back resource [53].

Other seasonal uses recorded are the use of wild plants for human food during the agricultural growing season [73]; and the prescribing of rest-days for gathering in the forest, so that the forest may have time to recuperate [20].

Management by area

Much woodland management is of course management by area, and the rules which define what the area is, and who the legitimate insiders and outsiders are, have been analysed in some detail in section 2. However, since there the emphasis was on ownership, and here we focus on management tools, the defined area is mentioned again here as one of those tools.

Management by taboo and religious sanction

Religious sanctions in the African context are a way of connecting elders or rulers to God and the ancestors, and using their authority to strengthen the authority of the living individual: of wrapping up management rules, in short. There are examples in this collection of the following sorts of sanctions:

- the attribution of calamities which occur to the illicit felling of trees among the Kikuyu [22].
- the threat that over-exploitation of the forest, or cultivating of over-steep slopes would call forth Assie, a female matrilineal deity who is dangerous to humans, among the Aouan of the Côte d'Ivoire [20].
- threat by the Hausa sultans of Zinder in Niger to decapitate anyone who felled a *Faidherbia albida* tree and to amputate the arm of anyone who mutilated a tree of this species [43].
- the creation of closed and open hunting or gathering seasons by the placing of religious taboos on infringements, such as the taboo

on the cutting of 'female' trees during the agricultural season in Botswana [88]

While we cannot always understand these taboos - partly because of the incredulous or superior way in which they are often reported, and partly perhaps because of the way in which the original informant explained the information it is clear that they were usually obeyed and that they had meaning for those who obeyed them. We should assume, therefore, that such sanctions are rooted in historical observation, and a good understanding of the local ecosystem.

For instance, it is not infrequently reported in some parts of Africa - much to the irritation of community forestry projects trying to get started - that villagers will say (as they did around Tamale in northern Ghana [70]) that anyone who plants trees will die; or (as they did in south-east Botswana [96]) that only God may plant trees and that it is impious for humans to do so. As bald statements, these can be dismissed as mere conservatism, or as proof that more publicity on the value of trees is called for. Closer investigation reveals in each case that the proposed tree-planting was on village CPR land, that the trees would diminish the CPR rights of others, and that it is other humans, not God, who may be wronged in due course. The speaker is simply telling the project that elaborate arrangements are under threat, in the strongest language he can muster.

Fire as a management tool

Fire is a very important management tool [99; 111], and is one on which there has been much conflict in the past between European foresters working in Africa and local people. Unfortunately the set of references gathered together here do not do justice to its importance, and it is probable that there is a need for further new field research on indigenous knowledge of management by fire. Two technical articles have been included to indicate what some of the conflict areas have been.

Fire is plainly both a cleanser and enricher of the earth, and a potent danger to savanna woodland. It is an essential part of agriculture, as well. The evidence would seem to be, however, that the Sahelian (and *miombo*) farmer has had a sophisticated understanding of the effect of fire on the tree and grass species he needed both for the presence of animals, and for bush fallow cultivation.

The negative tone of much professional comment on Sahelian herders' fire management practices can be read in [78] where fire-risky lopping is reported. We do not know whether the observer misunderstood what he saw, or whether indeed this particular herdsman was off his home range and treating the property of others with scant regard as is certainly common [69]. Yet elsewhere we hear of farmers planting firebreaks of bamboo in their agricultural system [38], and of a most efficient system for the policing of careless fire-users [85]. Here, on the left bank of the river Niger, the Fulani would put out bush fires they came across and, once the fire was under control, identify the culprit(s) and impose a fine: the provision of a feast for all those men who had put out the fire.

Two fire control experiments are reported on:
The first [26] was conducted in savanna woodland close to, and derived from, closed forest in Nigeria, over a 28 year period from the 1930s to the 1960s, and with plots which had been treated with different management methods. After 28 years:

- In the fire protection plot, savanna grasses had been eliminated, and 46% of tree stems were of fire-sensitive rain-forest species.
- The early burnt plots were going in the same direction,but more slowly.
- The late burnt plots had remained as open savanna woodland.

In the second, [72] experiments were conducted in *Anogeissus-Combretum* forest in Nigeria, (600-900mm of rainfall) on a variety of combinations of protection and burning regimes, to discover which gave the greatest mean annual increment, and which the greatest number of large stems per hectare.

1 By early burning every fourth year, and by fire protection the rest of the time.
2 By early burning and fire protection every alternating year.
3 By fire protection alone.
4/5 By early or late burning annually.

In order to obtain a large number of large stems per hectare (50cm and over), protection without burning was the best regime, even though some mean annual increment (MAI) was sacrificed.

These experiments suggest two things. In the first experiment, fire protection and early burning (which is less radical than late burning and so is a form of modest fire protection) were good ways of changing the savanna back into rain forest, but the last thing that a Sahelian herder would want. For him, late hot burns would apparently create the open woodland he wanted: the very thing he was already doing, and which had so incensed colonial period foresters.

In the second, a combination of burning and protection gives the best MAI, while protection only gives timber trees, but less MAI. Again, the Sahelian farmer-herder is vindicated here for his perception that burning enhances MAI.

Animals, grazing and browsing as management tools

The very range of vegetation to be found in the semi-arid areas of Africa is to an unknown extent the result of the planned and unplanned impact of grazing and browsing animals. The mean annual increment of woody vegetation can be increased by the reduction of herbaceous competition offered by grazing animals, so long as there is enough rain for it to grow at all [16]. Grazing also keeps the balance between annual and perennial grasses, and between grassland and bush, at the same time reducing the incidence of inopportune fires [111].

Africa's pastoralists are skilled at managing a broadly spread, light exploitation of plant cover by keeping a wide range of different stock with different grazing and browsing needs [67].

As made clear above, short fallow systems are fundamentally dependent on animals and their manure, while *Faidherbia albida* enables the farmer to keep at least some of his herd on the land through the dry season [74; 80].

Management of the individual tree

Mariam Niamir [69] in her survey, noted few published details of management of individual trees. Raison [84], while describing the pollarding and removal of side-branches practised on trees on terraces among the Mafa of Mandara, Cameroon, also comments on the lack of single tree management examples in the literature survey he conducted. This survey has turned up a small clutch of examples, however.

Lopping and pruning. The Kamba in Kenya lop side-branches from *Grevillea robusta* and *Eucalyptus spp.* in order to produce a central pole for construction timber in the long run, and some fuelwood in the short run [35]. Many Africans are aware of the species which respond best to lopping and which will readily sprout again, and plant and preserve them where they can [73].

A case in Sierra Leone is cited by Hoskins [49] of trees being cut to different heights when a fallowed field was being cleared in order to favour selective regrowth. In *Faidherbia albida* fallows, all the *F. albida* trees are lopped on a rotational basis every 3-5 years and fed to livestock, while the wood is used for fencing and fuel. The trees are thus kept to a manageable size in the field [51]. Summary [73] presents an interesting if puzzling management example. The author notes thorn-fenced fields containing heavily lopped trees, and fields surrounded by living hedges (thus more permanent) in which the trees are carefully and lightly lopped. Tenure difference would seem to be at work.

Pollarding. In Ghana, trees on swidden-fallows are pollarded so that when they sprout again the shoots will be above grazing height [4]. Pollarding is noted among the Mbeere in Kenya by Castro and Brokensha [24]. Poulsen [81] writes similarly of the unusual variant on pollarding practised on Mount Kenya where *Grevillea robusta* is heavily pruned every two or three years for fuelwood and small timber, while a larger pole can be grown centrally. The trees can be pruned for 30-50 years.

The Turkana climb high into fodder trees to pollard them when selecting dry season fodder [9] and this practice is widely reported by ILCA for the subhumid zone just before the rains when grass is no longer available [53].

Coppicing. In some silvo-pastoral systems, trees are cut to stump level for various purposes (timber, fuel, browse) and then left untouched for several years in a swidden-fallow period [67].

The tools for the job. The usual tool for all these actions is a bush-knife - a panga - or in some cases a small axe. Only commercial fuelwood traders use heavier tools such as long-handled axes or chain-saws [96].

Individual tree-management: conclusions. Many of these examples seem to show management of the individual tree so as to diversify its products and its usefulness, and to make it work harder for its keep, as it were. More intensive tree-management is found mainly on farms, presumably because while an ample bush supply still exists, diverse products will be met from diverse trees rather than by management of one species alone.

5. Woodland management and the state

One of the facts which emerges from the summaries gathered here is the tremendous paucity of formal forester knowledge about the management of tropical dry forest, yet of the enormous changes wrought upon them in the name of better management and State control.

Key survey articles from the last few years hammer the first point home. According to Bonkoungou and Catinot [18], there is only very slight experience of silvopastoral management based on natural regeneration techniques for mixed forest and grassland. Much indigenous knowledge waits to be collected. According to Jackson [54; 55], who did the biggest-ever survey of the Sahelian literature on woodland management, there were only two examples, up to 1983, of formal forest management in the Sahel: that of Bandin forest, Senegal (which failed in its aims) and the relatively successful management of *Acacia nilotica* on the Blue Nile in the Sudan. Other projects failed because they failed to build on the much longer technical experience of local people, (e.g. [82] a windbreak project in Cameroon).

Set against this thin knowledge, we find the imposition of European concepts of property and land-tenure, with disastrous effect. The most important gap was the failure to understand the swidden-fallowing systems, which had used the landscape sustainably for hundreds of years. The systems were simply invisible to outsiders [93] and land being fallowed often looked abandoned and ownerless to the northern forester's eye. As a result, fallows, rested for years, well-wooded, and almost ready to be felled again, were gazetted by the state and turned into forest reserves (forets classés), turning intensively managed CPRs at a stroke into open access land as far as villagers were concerned, to be poached from rather than managed [49; 110]. People became wary of fallowing their land [103] and overworked it instead. In this selection of summaries, instances of State appropriation of CPRs and fallows in this way occur in no less than eighteen cases [5; 11; 15; 27; 28; 32; 40; 49; 62; 80; 90; 93; 94; 97; 98; 103; 105; 110].

There is a further confusion. In the literature which deals with francophone west Africa [e.g. 99], land tenure and tree tenure are often described as if they are two distinct entities, and have been so since pre-colonial times. The anglophone literature on the other hand seems to see the two as always going together, [e.g. 11; 12; 35; 86] even while it is made clear that people do have limited usufruct rights in trees on each others' land in the dry season or while it is being fallowed. The apparent difference is resolved if we remember that land farmed by individual farmers was itself only 'borrowed' from the lineage to which, when not under annual crops, it returned, along with any trees growing on it. Even the association of individual valuable tree-species with the local ruler was no more than an expression of this communal tenure. However, in the opinion of Norton [71] and this author, the separation of tree and land tenure became a reality in some countries only in the colonial period - as the result of poor understanding of fallowing rules. Indirect support is lent for this view by Thomson's work on Niger [102], and by Seif el Din's [90] on the Western Sudan. The assumption of a historical split between land and tree tenure has been used to reserve certain tree species for the State to manage, to the ridiculous point where it is as impossible to fell a *Faidherbia albida* tree today in Senegal [80] as it was under the ferocious

Sultan Tanimoun in 19th century Niger [25], (who also seemed to misunderstand the nature of fallowing systems). State management has thus not only lacked insight, it has eroded working systems and concepts to a serious degree.

6. Woodland management and other change

Tenure and land use changes

Villagers have been turned from land owners into leaseholders in many countries e.g. Senegal [80], or Sudan [90]. Or indeed from land owners to landless wanderers in the case of pastoralists. Barrow points out that customary tenure was taken into account in Kenya when agricultural land was demarcated, but neither investigated nor recorded in the dry areas [9]. The best solution offered to Kenya pastoralists has been group ranches, which lack the area flexibility of lineage range management [76] and begin to look like some of Africa's other tenure ghettos, such as the Communal lands in Zimbabwe [106].

Political changes affecting land tenure usually destroy CPRs. For instance, during the Ujamaa period, the Sukuma were asked to leave their dispersed granite outcrops and to cluster in villages. Unoccupied, these areas fell prey to urban charcoal burners [95].

Land-use changes relating to economic change

The introduction of the peanut has had a negative effect on the value of oil-bearing tree-species such as *Butyrospermum parkii*, and the systems into which they used to be set [84] and has also led to excessive clearing of forests in many areas, [30]. On the other hand, it is a crop which fits well into curtailed swidden-fallow-cycles.

The loss of authority of the elders

In Turkana, the authority of the clan elders, who were originally solely responsible for livestock and natural resource management, is being eroded by modern education for the young, and by the presence of government officials in the area. Among the Kikuyu, as long ago as the 1930s, the weakened position of the elders made the institution of sacred groves no longer tenable [22].

Population growth

Better health conditions and food security for humans, and the increased use of veterinary drugs, has increased both human and animal pressure on rangeland and farmland. More people are living in fixed, rather than semi-mobile conditions, with a corresponding heavy intensification of land-use and biomass offtake. These increases have not been incorporated into land management strategies even though the range in dry areas was already used at close to its subsistence potential [9; 103].

Many parts of Africa have experienced bigger influxes of people than they can easily absorb because of those moving southwards away from drought areas [45], because of refugee problems [111], and because of the artificial withdrawal of large tracts of land, as in the mechanised farming schemes of the Central Sudan [111].

As populations rise, increased sedentarisation occurs [74], the household grows in importance while the clan and lineage structures shrivel [73], and grazing is privatised [12] or shared by a much smaller subsection of a larger entity [34]. All such actions sound the death-knell of successful woodland management.

It is a truism that rising population pressure on forests will cause deforestation. For areas with sufficient rainfall, the inevitable longterm result of population pressure is in fact a move to tree-**planting**. In Rwanda [40], a survey revealed that people plant trees most where there is the most pressure on land and much cash-cropping. Male farmers who got fuelwood from their own land had planted more than those who collected dead wood or whose wives got the fuelwood.

The shortening of fallows

Population pressure seems to have led to the shortening of fallow periods throughout semi-arid Africa [2;15;31;89;90;109]. In the western Sudan the whole cycle has more than halved in length, and the fallow period is less than a third of what it was [46]. The tendency is for the fallowing system to shrink in the end to the point where it is replaced by a crop rotation of alternating millet and peanuts [84].

The concentric farmed parklands are under other kinds of pressure too. If they cannot be abandoned at the end of their cycle because of overall population densities, they degrade, and tree-stocks fall to perhaps only 5 trees per ha [80;84]. Labour migration from the Sahel has encouraged extensification of agricultural techniques, for although the population overall has grown, more men and more young people have left the area for ever. Instead people are trying to cultivate the most land they can in the least time, at the very beginning of the rainy season. Each household head now tries to spread his bets by sowing over as wide and varied an area as possible. The orderly management of land practised in the past has tended to fall apart, and more and more marginal lands are farmed, with accompanying destruction of bush areas. Older practices such as manuring, intensive sowing and weeding, planned swidden-fallowing and water conservation, have all been replaced by quick easy farming [103].

The growth of towns

The growth of towns is one of the largest threats to the dry tropical forests, because the clustering of large numbers of people who will need rural biomass for fuel, and the low MAI of that rural biomass, mean that the fuelwood shadow from a large town or city spreads hundreds of km out into already hard-pressed rural areas. Yet population growth in the Sahel is of the order of 2% or less in rural areas and 6-8-10% in urban areas. In addition, urban entrepreneurs are keeping cattle for cash near towns, where before they would have travelled widely with them [15; 16; 29; 30; 45; 59].

Roads are built from towns to rural areas, and these then facilitate more commercial transactions, such as the sale of charcoal or wood-cured tobacco to towns [24]. Labour migration to towns leaves rural economies unable to farm properly, and leads to the extensification noted above. Furthermore, the World Bank now believes that urban-biased economic policies bear responsibility for further extensive agricultural production and the shortening of fallows, because of low producer prices [109].

Thus towns grow at the expense of rural people, shrinking their resource base and extracting huge quantities of precious biomass from them [94]. Urban people can make rural resources come to them in effect, and can even be absentee herders, enjoying the wealth animals bring without having to take care of the resource the animals live from in the way true pastoralists do [107].

The State itself, as an urban-based institution, has exploited but failed to understand the dynamics of rural systems, has seen its own urban needs as paramount, and in sum has made it harder for rural people to build a surplus of animals, of grain (because of labour shortages), or of fallowing land. The result has been much increased vulnerability to cyclical droughts, let alone the more substantial dry phase we are witnessing at the moment.

Climate

Finally, and beyond human control, it is clear that the last twenty years have been exceptionally dry in semi-arid Africa in contrast to the period which immediately preceded it. Up to 1968, livestock raising increased steadily and natural bush regeneration was abundant. Since 1968, however, natural regeneration has not been favourable and woodlands are changing their species composition, and ageing. In some places, as rainfall decreases, we see species changes such as *Acacia tortilis* taking over from *Acacia laeta* and *Acacia senegal* [19]; in others, *Acacia tortilis* and *Commiphora africana* have been dying, woodlands have opened up and erosion has set in. Also, by the 70s, gum stands from the 40s and 50s had almost disappeared and no regeneration was present [33].

Groundwater deficits have stopped tree species like the *Butyrospermum parkii* and *Parkia biglobosa* from regenerating [16], and have caused the Niger river to drop by 50-60cm, with immense recharge deficits as a result [56].

The result is that a forest like that of Rawashda in the eastern Sudan [107], originally reserved for urban fuelwood needs, would actually now need to be much larger to satisfy the same needs as before, because of reduced MAI. But the whole area is already under much more pressure than it was when it was reserved.

7. Four participatory woodland management projects: Guesselbodi, Niger; Rawashda forest, Eastern Sudan; Bay Region, Somalia; and the Turkana rural development project, Kenya.

All of these projects have attempted to involve local people in forest management of various kinds, and all are documented in the attached summaries. Guesselbodi appears in [47] and [59]; Rawashda in [97] and [107]; the Bay region project in [17] and [94]; and Turkana in [6;7;8;9] and [60].

project rationale

Guesselbodi and Rawashda are forest reserves set aside for the provision of fuelwood to an urban area. Both had lost a good deal of their tree cover over the years, both were used in all kinds of ways by local people (especially for grazing and as livestock routes) and a participatory approach was conceived of as a way of limiting the local threat to the resource by offering employment and slightly improved rights to local users.

The Bay Region project began as a forest inventory project in the area from which Mogadishu's chief charcoal supplies came. Only in the course of the project did it become clear that there were serious conflicts between the herders in the area the charcoal was coming from and the State charcoal co-operative, **and** that the remaining potential charcoal supply was quite modest. Work was begun on a phase 2 which would pass substantial management of the resource to local people, and at the same time would lighten the burden of charcoal offtake by addressing fuel substitution in Mogadishu.

The Turkana project was begun with the intention of helping the people of Turkana recover from the droughts of the 1980s. Its focus was always local.

area identification

Guesselbodi and Rawashda are forest reserves, gazetted many years ago. The projects are making an attempt to graft local involvement and management onto a pre-delineated piece of land. In the case of Bay Region, preliminary work was done (before the Somali civil war made the region a no-go area) to identify the interests of local people in forest management and to allow them to plot the boundaries of the areas they would manage, following earlier indigenous land divisions. The Turkana project is involved in many aspects of people's lives, but as far as forestry goes, it too has identified the areas with meaning to the local people, and is building on existing management structures to enhance tree-protection in the Region.

the interest of local people in a management role

At Rawashda, local people at the seminar reported by Vink [107], said that they were keen for the benefits of the forest to go more to local people, and wished to see charcoal burners dealt with severely. They did not want to take part in management, since they saw that as the Forestry Department's job, and were reluctant to enter into **taungya** arrangements.

At Guesselbodi, local interest in the project was minimal until it became clear that there would be financial benefits from the fuelwood co-operative to which villagers could sell the wood they cut, and there was then much more enthusiasm. But villagers are not prepared to do any unpaid management work until ownership and control questions are settled.

In Turkana and in the Bay region, individuals are already involved in aspects of forest management, and see the respective projects as ways of enhancing their capability.

attempts to understand indigenous tenure

The Guesselbodi project spent almost two years on questionnaires about forest users and forest uses before management began - but we hear not a word about tenure except the statement that there are difficulties in implementing the French Loi Forestiere because it alienates people. At Rawashda, the consultants' report of 1987 [97] points out that there is conflict locally over tenure in the forest which is seen historically as tribal CPR land alienated by the government when the reserve was created. There has certainly been no move to de-reserve the land during the last three years of the project, nor (to my knowledge) has the reserve ever been redesignated a local reserve rather than a Central reserve for the supply of city firewood. Both the Turkana and the Bay Region project, by contrast, take indigenous tenure rules and areas as their starting point as far as possible.

authority structures

Predictably Guesselbodi and Rawashda start with the assumption that the project has the authority to decide on management patterns and to fit villagers into that. Its is unclear what institutions in the Sudan are being used, but fortunately for the project, village sheikhs are so powerful in that country that they can deliver an actively involved village if they wish to. The seminar held in 1986, at which alternative management plans for the forest were discussed with local people, and their views taken, was an excellent event. At Guesselbodi, the management plan was presented to the people when it was already finished. No attempt to investigate local authority structures was apparently made, and a cooperative was created for handling wood purchase and sale. By 1990 [59] it was becoming apparent that a cooperative 'may not be financially or socially the best institution for local participation in forest management'. In the Turkana and Bay projects, the project works with, or would work with, already respected local leaders and chiefs, and build on their management strengths and capabilities.

lessons to be drawn from these cases

It would seem that a project which starts with a piece of forest whose boundaries have been defined under one management regime, and which then looks for local management inputs, will have many more difficulties at every stage than a project where the need and interest of local people is the starting point, and forest management emerges from that commitment.

The area to be managed has to be commensurate with the management capabilities of the target local group; that it why it is so essential that local people help to draw up boundaries within which they feel competent. If a forest is essentially to change owners it will have to be subdivided into smaller management units.

To put it another way, forest management projects where the real intention is to protect the forest by attempting to buy off negative local use with some local involvement (but no rights), are unlikely to succeed. As one of the officials with a responsibility for Rawashda said in conversation, 'People are not *that* stupid'. Reserved forests for **non-local use** are particularly poor candidates for local management - since by definition these are reserved **from** local people essentially.

Early involvement in **decision-making and planning** is obviously crucial, and a relationship of equality between those who run the project and the local authority figures with whom they will work. A long-term relationship will have to be forged, of the kind that has been managed so well in Turkana.

8. Conclusions

The nature of indigenous management

Diverse as all the management examples gathered here are, certain themes stand out. Firstly, in the past, there have been strong capable managers in charge of woodland management and the exploitation of trees, in many of these areas; managers with a lifetime commitment to the area, who are making their living from the resource they are managing just like the people they administer, and who are often related to some or all of the people on whose behalf they issue management rules. Most management rules, as a result, are very well attuned to local needs and constraints, and have arisen in apt response to some perceived problem [21].

Secondly, management is as simple as possible. Unless the resource has some value or some scarcity, management will not be undertaken [13; 40; 77]. Any management rules which are devised are quite flexible, and can be modified as need arises.

Thirdly, management is for a set of interlocking benefits. It is quite hard to separate out **woodland** management from swidden-fallow management, herd management, and annual crop management. Moreover wood is far from being the only resource for which woodlands are managed [15; and passim].

Fourthly, rising population density is turning pastoralists into farmers, long swidden-fallow into short, the usufruct of clan land into individual title. So the management focus has narrowed and in many areas the numbers of locally born and locally significant decision-makers above the level of household head are dwindling.

Finally, political and economic authority has passed from indigenous managers to the State in most places. The elders left can no longer command the respect they used to, and so it is difficult for them to hope to manage forests or woodlands in any very complex way.

Management and change

Management has changed from use-rights based on clan-membership and thence the right to use clan resources, to the exercise of State-granted privileges and management by restriction and exclusion. The insiders have become outsiders. Similarly, rangeland and pastoral systems are under stress or in a state of collapse in many countries and areas [69].

Involvement in the wider political economy undermined the political authority of the kin-group and allowed an increasingly unregulated exploitation of land. Unfortunately, these changes occurred during the abnormally wet 1950-1965 period, so that the initial effect of negative resource management changes were cushioned by good rains. The 1968-73 drought exposed the breakdown in these kin structures and their careful resource management practices.

Yet centralised political authorities continue to deny, on the whole, the ability of local decision-making bodies to manage their environment; and government legislation has become necessary for the smallest changes to established practice, dissuading groups from organizing [109].

Because local laws are not in effect any more, degradation within the forests has resulted, with excessive lopping and pruning and ageing of the trees [25]. Forestry has come to mean forest reserves and village forestry schemes, neither of which replicate the integration with trees practised in the past. [92].

The prognosis for adaptive change looks poor, from many points of view. Many previous woodland management practices are only likely to work under conditions of low population density, and can certainly only work if managers are also owners. Yet it is rare for reserved forest land to be returned to the people who are being asked to co-operate in its management.

There is also the question of its size. It would seem that rural people only want to manage a resource from which rule-of-thumb estimates suggest that all can benefit to some extent. Small patches of hilltop forest can be managed by small numbers of people, or made into a sacred grove if there is a political institution which can guard it. Otherwise, rural people would often prefer such small patches to be looked after by the Forestry department, so that they are not involved in high social costs for a low return.

Ironically, of course, tree management increases the less forest there is. Under particular circumstances trees are managed as forest, but increasingly preferred management is on farms (e.g. [108]). A publicly owned forest is an anachronism in a tightly farmed landscape, and often represents the superimposition of the will of outsiders on the local population.

Lessons for possible projects

Many of the management techniques discussed in this document are too embedded in particularity to build projects on. Of those which remain a possibility, ownership of the resource - either by descent or by residence, so long as all parties agree to the principle - is fundamental. No management will take place without it. The recognition of ownership in a public way - in written contracts, by letting local management preferences and skills help to shape plans, etc. - is also a sine qua non.

The importance of the investment of (unpaid) labour is often overlooked as a creator and maintainer of tenure. The ownership credit in the case of paid labour goes only to the payer, of course. The willingness to provide some unpaid labour is a sign, as well, that those taking part in management actions do so because they see benefits for themselves by their actions.

Management by area will be the other most commonly used management technique. Given the tendency for large earlier collectivities to break up into smaller groups, smaller areas for management may well have to be found. The Mwanza hilltops before the Ujamaa period are a good example of a small resource, managed by a small number of neighbouring families [95].

The case examples suggest strongly that local resource management stands a better chance of success somewhere relatively remote (such as Bay Region) than within the catchment of a town, like Guesselbodi or Rawashda. Remoteness, a low population density, and often relatively low rainfall, are the keys to areas in which successful local woodland management might take place.

In nearly all other cases, the most promising focus for local people is on the creation of tree resources on the farm, leaving patches of environmental reserve to the State - the successor of the leaders who managed sacred groves - or in rare cases to small local user groups whose internal composition and whose managed resource are clearly defined and felt by them to be manageable.

When institutions of the strength of the clans and lineages break down, institutions which have allocated land for hundreds of years, we should not lightly suppose that we can create new arbitrary village institutions to do their work.

Suggestions for further research

Three woodland management research topics emerge from this African survey.

Firstly, experimentation is clearly needed in the devising of land-use systems which combine the management of off-farm tree-resources shared by small groups, with normal on-farm tasks. 'Small CPRs' are probably going to be an important new direction for research to take, and they need to be combined with work already done elsewhere (e.g. Nepal) on the identification and support of user groups. Because tenure modifications are needed in many places to make this possible, legal research must accompany CPR research in most cases.

Secondly, what should be our approach to all the forest reserves still to be found in Africa? Which should be retained by the State and why? (Those with only diffuse and indirect local benefits, such as those to be retained for watershed or biodiversity reasons? Those intended for non-local fuelwood supplies to big urban centres?) Which should be broken up and allowed to become farmland, and which might best be managed by local people as woodland? No new thought on the subject of reserves has taken place anywhere for decades in Africa, yet population densities have risen and Forest Departments are poverty stricken.

Thirdly, there is much data still to be collected on indigenous knowledge of fire as a management tool.

Finally, and worryingly, it is clear that there is an unnerving tendency in the literature scanned to see local management as a cheap option for financially hard-pressed forestry departments, in which control and ownership is retained by the State yet local people - mysteriously prepared to do so for no pay - do most of the management because of their needs from the forest. This approach must be resisted for two reasons. **Firstly**, it is exploitative: or it would be if it were not for the case that, **secondly**, it does not usually work. The summaries here make it clear that management practices are only undertaken by local people if the benefits outweigh the costs. If it is simpler to leave the State in charge and to steal the benefits needed from the forest, that is what will take place.

Thus it is important to stress, in offering a literature review of this kind, that the author has no intention of accidentally endorsing the weak version of participatory forest management (in which no strengthened tenure rights are envisaged). No collation of materials of this kind should be taken to suggest that local people can be made to act economically irrationally off their farms in a way they would never do on their own land.

Part II
The Bibliographic Summaries

1. Abrahams, R. G., 1967. *The Peoples of Greater Unyamwezi, Tanzania* Ethnographic Survey of Africa, D. Forde (ed), East Central Africa, part 17, International African Institute, London.
TANZANIA This summary concerns West Central Tanzania, where the people practise agriculture, livestock raising, and some hunting in the dry season. Men trade in honey, beeswax, ivory and dried and fresh fish. Land belongs to the chief, and through him to the lineage members. This applies to both farm land and to the land, trees, wildlife and other products of the bush. Management is by area and by lineage membership. Strangers have to ask permission before making use of land or any other asset in the area.

2. Ahmed, A. A., 1983. 'Forest Reserves and Woodland Savanna Regeneration on the Sub-Saharan Massif of Jebel Marra, Democratic Republic of Sudan (Natural Forest ecosystem recovery)' in *Vegetatio* 54 (2) pp.65-78. The Hague: W.Junk.
SUDAN Jebel Marra, Sudan is situated in the Sahelian zone, but the high massif modifies the micro-climate and increases precipitation; the highest rainfall is 1000mm per annum on the western slopes, where temperatures are also low, and the area is protected from dry northeasterly winds. Average precipitation is 600-1000mm. The vegetation comprises mesophytic woody species and mesophytic woody plants. The main tree cover is *Anogeissus leiocarpus, Cordia abyssinica, Sterculia setigera, Lonchocarpus laxiflorus, Khaya senegalensis, Faidherbia albida*, and *Ficus palmata*.
The inhabitants are sedentary cultivators practising swidden-fallowing[1], and covering all the accessible slopes of the massif from the base to 2750m. In Jebel Marra, a bush swidden-fallow system is operated. Plots

---

1. The hybrid term swidden-fallow has been used for two reasons. Firstly, it links two terms which, rather arbitrarily, are normally used respectively for non-African and African situations. Secondly, it stresses (in a way terms such as 'shifting cultivation' and 'slash-and-burn agriculture' do not), the fact that farmers do not abandon cleared land when they leave it to restore its fertility and plant crops on new land. They obtain other products from it, may retain continuing control over trees growing up on it, and will return to it when their own particular cycle is complete.

are cropped for 3-7 years, then left fallow for up to 15 years. In response to population pressure, swidden-fallow periods have been progressively shortened.

3. Allsebrook, D.L., 1987. *Learning to combat desertification in the Sahel*, MSc thesis, Reading University.
REGIONAL The author's focus is the Sahel with its fragile soils, intermittent rainfall, and ecological diversity. The inhabitants are sedentary cultivators and transhumant pastoralists, and trees are used for food, medicines, handicrafts and fuelwood. It is suggested that common property ownership by powerful clans is an important component in the prevention of ecological destruction, as the clan may enforce punitive measures by writ over a far larger territory than it is possible under individual ownership, which pertains to small areas around homesteads. In many Sahelian rural communities, religious groves 'revered forests', 'ghost trees' etc. constitute a religious symbolism which is an informal and indirect means of regulating access to firewood sources.

Indigenous resource management practices are not fixed, but are constantly undergoing change as on-farm environmental problems arise. They have the potential to find solutions to the problem of sustaining the resource base. However, Sahelian land managers may be constrained by desperate short-term needs for food, fuel, security and money which force them to adopt more short-term destructive strategies. The inability any longer to implement long-term strategies is attributed to the breakdown in indigenous social and political structures.

4. Ardayfio E. 1985. *The rural energy crisis in Ghana: Its implications for women's work and household survival*, WEP Research Working Papers WEP 10/WP.39, ILO, Geneva.
GHANA The research is set in the southern forested area, where the main activity is agriculture, using plough animals. Land rights are invested in the lineage or stool and any member of the lineage is entitled to use any land which is lying fallow, and it belongs to him or her for as long as he or she uses or occupies it. Trees from the owner's own swidden-fallow are pollarded for fuelwood, so that cattle cannot damage the new shoots.

5. Barnes C., 1984. 'The historical context of the fuelwood situation in Kisii district', pps. 61-78 in *Wood, energy and households* Barnes C., Ensminger J. and O'Keefe P. (eds), Beijer Institute, Sweden and Scandinavian Institute of African Studies, Sweden, 1984.
KENYA The Gusii people live in the southeast of Kisii district, Kenya.

The region ranges from 1500-2350m elevation and contains many permanent streams and some swamps. Rainfall is 1500mm per year. The population has increased from 122/sq km in 1948 to 395/sq km in 1979. Indigenous trees grow only in river valleys and isolated areas. The Gusii are settled agriculturalists growing cash and subsistence crops with some livestock. Coffee and tea are the main cash crops; sugar cane and bananas are also sold.

In the past, clan land was divided into settlements with arable fields. Each neighbourhood had a series of common fields comprising individual plots, and a bush area used for common pasture. Communal areas now belong to the State and most land comprises permanent private plots. Under communal land systems wood was freely available for collection and cutting to members of the clan. With privatisation of land holdings households are able to obtain their wood needs from their own land; through agreement with a landowner where there is wood available; or from Crown land, where it may be gathered but not cut without the permission of local government. Thus management has changed from use-rights based on clan-membership, to the exercise of State-granted privileges.

6. Barrow, E.G.C., 1986a. *Value of traditional knowledge in present day soil conservation practice, the example of the Pokot and the Turkana.* Paper presented to the Third National Workshop on Soil and Water Conservation, Kenya, 1986. Example 1 The Pokot.

KENYA Nginyang Division, Kenya, is in ecozones marginal and semi-arid. Here, live a transhumant pastoral section of the Pokot tribe with goats, sheep, cattle and camels. Trees are used sustainably for a variety of uses including fodder, food, medicines, building materials, fuel, fencing, shade, household implements and as the venue for elders' meetings.

Within the broad grazing pattern areas are set aside for reserved grazing, usually in the form of reserved and guarded hills. The reserved grazing is used at the discretion of the elders during the dry season or drought years. The Pokot attach great value to trees (both materially and culturally) and will rarely cut a valuable tree. During the dry season some trees (e.g. *Balanites aegyptiaca, Dobera glabra*) will be pollarded for browse and pods will be harvested for livestock feed. The only woody species that are actually cut back are the less useful bush species (e.g. *Acacia reficiens* and *A. brevispica*) which are used for fencing. This cutting back of such bushy woody species often serves to encourage a better ground cover of perennial grass. There is considerable knowledge of the flora, especially in relation to animal fodder. Woody species can be recognised that will promote milk or meat production, dry and wet season fodder, fodder for different stock species and ages.

7. Barrow, E.G.C., 1986b. *Value of traditional knowledge in present day soil conservation practice, the example of the Pokot and the Turkana.* Paper presented to the Third National Workshop on Soil and Water Conservation, Kenya, 1986. Example 2 The Turkana. (With a little supporting additional data from Barrow, E.G.C. 1988. *Trees, people and the dry lands: the role of local knowledge.* Institutional invited paper presented to the Second Kenya National Seminar on Agroforestry, Nairobi, Kenya.)

KENYA The Turkana, Turkana District Kenya, live in a semi-arid/arid environment. They herd camels, cattle and goats and also have sorghum gardens. In a good wet season, forage is plentiful and so there are no restrictions on stock movements and questions of resource ownership do not arise. However in the dry season and in dry years such issues become critical. It is this power over ownership of water and fodder that is central to the Turkana. Dry season grazing takes place in higher rainfall areas, and is particularly critical in the dry plains area. Here sublineage ownership of riverine woodland, containing important fodder tree species such as Acacia tortilis, is well defined. A herd owner's home range or dry season base is his *Ere*. Close kin have long associations with such areas and their ancestors are often buried there. There, they have ownership rights to resources which include fodder and fruit trees and a dry season well. Outsiders will not be allowed to use these resources without prior permission. Within the larger group *Ere* area, sub-group resources known as *Ekwar*, (meaning 'trees by the side of the river'), include the right to exploit important individual trees or small stands of trees, as a 'private' dry season fodder and browse reserve. Individuals have to have a strong network of close kin and supporters to keep *Ekwar* rights intact over time, and at times the elders have to adjudicate between rival claimants. There are individual user rights over certain important species : *Acacia tortilis* for fodder, *Hyphaena coriacea* for basketry, *Dobera glabra* for fodder and famine food, and also over *Cordia sinensis*, *Zizyphus mauritiana*, and *Faidherbia albida*).

The Turkana have a well developed indigenous knowledge of their flora and its uses, with especially detailed knowledge of the browse potential for different stock, and in different seasons, of local tree species. Very few trees will be cut back completely - pollarding or lopping branches is the norm. Trees are used for construction, household utensils, fodder and medicine.

Individual user rights are only recognised as long as the families concerned continue to exercise them effectively at harvest time, and rights lapse if they are not maintained. Originally, in the Turkana riverine sorghum plots, trees were not generally cut down, and only the bush and undergrowth was removed. Yet now in the areas surrounding the riverine irrigation schemes (which have trespassed on the Turkana *Ere*) all the rainfed crops are cleared of trees, in imitation of the irrigation scheme.

8. Barrow, E.G.C., 1987. *Extension and learning examples for the Pokot and Turkana, pastoralists in Kenya.* IDS workshop, Farmers and agriculture research: complementary methods. Institute of Development Studies, Sussex, England.

KENYA The summary concerns Turkana district, a semi-arid and arid environment with high ambient temperatures and low and erratic rainfall (180-400mm). The inhabitants were originally highly mobile herders of camels, cattle and goats, but now settlements have been established all over the district. Trees are vital for fuel, construction, fodder, food and medicine. Woody vegetation is extremely important as dry season forage. The *Ekwar* is a vital part of livestock management: in the dry season it is often access to fruit and fodder trees that restricts movement. The *Ekwar* system is less strong in the wetter areas where there are fewer rivers with riverine woodlands. Preliminary analysis indicates that people are based in their *Ekwar* for much longer periods than planners associate with pastoral peoples. Usually at least one generation has stayed in their *Ekwar* and their fathers will have, in many cases, been buried there.

At the present the woody resource of the area is not seriously depleted except in the areas of settlement and in particular the larger settlements. However there is an overall lack of recruitment of young trees to the woody resource. The Forest Department and NORAD have run training courses on woodland management. Permits have to be obtained for burning of charcoal (and then only from identified naturally dead trees) and for building timber. This is enforced, in co-operation with the Administration, through the Chiefs' Act and is gradually achieving a degree of order in a difficult situation. These efforts have resulted in more people planting trees around their homes, without receiving any form of payment. There is also a much greater respect for the protection of young naturally growing trees and as a result many of these young trees are growing rapidly. Much of this is taking place in and around the settlement areas which is where the major problems are.

9. Barrow, E. 1988. 'Trees and pastoralists: the case of the Pokot and Turkana',*ODI Social Forestry Network Papers*, 6b.

KENYA The Turkana manage their woody vegetation on a sustained basis, selectively pollarding, rather than felling, valuable trees, using dry and dead timber for fuel, and less useful bush species for fencing manyattas (cattle kraals). Better health conditions and food security for humans, and the increased use of veterinary drugs, has increased both human and animal pressure on the rangelands. But these facts have not been incorporated into land management strategies even though the range was already used at close to its subsistence potential. More people are living in fixed, rather than semi-mobile conditions, with a corresponding heavy intensification of land-use and biomass offtake. The authority of the clan elders, who have in the past been responsible for livestock and natural resource management, is being eroded by modern education for the young, and by postings of government officials to the Turkana area. In high potential areas, indigenous land rights are recognised during the process of land demarcation, but they are neither investigated nor recorded in the dry areas.

10. Beaton A.C. 1948. 'The Fur' *Sudan Notes and Records* 29(1):1-39
SUDAN The Fur are Muslims who live in the western district of Darfur, in Sudan. They occupy the high backbone of land across the Nile-Chad watershed. Rainfall averages 555 to 650mm annually. The region is well forested with *Acacia spp.*, especially *Faidherbia albida*. Other common tree species are *Commiphora spp., Zizyphus spp., Combretum spp.*, mahogany and bamboo. The Fur are settled agriculturalists who keep some livestock and who market salt. Crops cultivated include wheat, maize, beans, sorghum, potatoes, vegetables, tobacco, groundnuts and cotton. Both sexes care for livestock. Wood is needed for fuelwood and construction. In addition, tree fruits are eaten which include *Balanites aegyptiaca, Cordia abyssinica, Ficus spp., Sclerocarya birrea* and *Tamarindus indica*. Whoever clears bush for cultivation gains rights to the land, and cutting trees is men's work only, (not least because women are the representatives of another lineage and the land cleared by men must clearly belong to the group to which he belongs - ed.)

11. Behnke, R.H., 1980. *The Herders of Cyrenaica*. University of Illinois Press, Urbana, Illinois.
LIBYA The book is concerned with the Bedouin in Cyrenaica, Libya, who live on the slopes of Jebel Akhdar, relatively well-watered above 800 metres, but fading to 200mm of highly unpredictable rainfall at altitudes of 200 metres and below. Tree cover is higher at higher altitudes, and sparse in the plains. A variety of different kinds of animals are herded.

Customary tenure (with which the text is in fact exclusively concerned) was calculated on a genealogical basis: all the descendants of one ancestor sharing one large area, within which more closely related kin shared smaller, closely adjacent portions. Thus any one piece of land had owners with strong claims and owners with weak claims to it. On the higher slopes, fruit trees such as apricot and almond were grown, while on the lower slopes scrub oak, *Juniperus sp.*, and finally *Ziziphus spina-christi* were browsed. In practice, farmed land was owned on a more or less private basis, while range land was shared: a practice which related to the irregularity of rainfall and the impermanent value from year to year of any one piece of land. In the areas of more stable productivity, high on the slopes of Mount Akhdar, access to land was restricted in various ways; in the plains, where productivity was erratic, access to resources was naturally much more open. It was also noticeable that labour inputs were what in part limited tenurial rights to more limited categories of people. On the mountain, where crops could be grown and water-points improved or created, such assets belonged to those who invested most, and most constant, labour in them. Thus increases in the reliable productivity of field sites were parallelled by increasingly exclusive terms of ownership. Trees fitted the tenure patterns of the land on which they were located - privately owned on the mountain side, and communally owned in the plains.

12. Behnke, R.H., 1985. *The dynamics of open-range management and property rights in Pastoral Africa*, ODI internal document. (Section 3 the Baggara of South Darfur)

SUDAN The Baggara, of South Darfur, Northern Sudan, are agro-pastoralists who manage their resources through the mechanism of the lineage and who herd cattle. The Baggara originally practised a form of customary tenure calculated on a genealogical basis: all the descendants of one ancestor sharing a large area for such purposes as grazing, but focusing on a small segment of the genealogy when more intensive land-use, such as agriculture, was planned. Thus any one piece of land was owned simultaneously by small numbers of people with strong claims, and large numbers of people with weak claims to it. Trees are important as browse, especially in the dry season and are owned by the owners of the land on which they stand.

The Baggara are in the process of privatising once-open range-land, claiming that they are doing no more than following original lineage rules, which allow for the intensification of ownership under certain conditions. In drier areas they do it by making sure that they control scarce water-points (and thus the grazing around them); in wetter areas they simply fence grazing areas and treat them as arable fields (which they may have been or may become). Such land-tenure changes are most common:

- where graze and browse are most under pressure from rising cattle numbers, or where there is a market for fodder (situations found near the town of Nyala where markets for milk and livestock on the hoof existed);
- where a particularly well-favoured grazing area is now needed at different points in the annual cycle by permanently resident local users, and by nomadic herders passing through. In both these situations, land had acquired a high value, and in consequence the range of 'legitimate' users has been narrowed by those with the strongest and most permanent interests in making sure that they retain access to the resource.

13. Beidelman, T.O. 1967. *The Matrilineal Peoples of Tanzania* Ethnographic Survey of Africa, D. Forde (ed), East Central Africa, part 16, International African Institute, London.

TANZANIA Much of Ukaguru, one of the tribal areas described in this survey, is covered in *miombo* woodland. Rainfall is around 750mm per annum.

The people are agriculturalists, without animals. Men collect honey and wax and women make a great deal of cash from honey beer brewing. Labour creates ownership and maintains it, particularly with regard to the clearance and cultivation of land. If ownership is abandoned, the resource reverts to the matrilineage. Management is work, and will not be undertaken unless there is the desire to own the resource to be managed.

14. Benneh, G., 1987. Land tenure and agroforestry land use systems in Ghana. *In*: Raintree, J.B. (ed.), *Land, trees and tenure.* ICRAF and Land Tenure Centre, Nairobi and Madison. 163-168.

GHANA The summary concerns the northern savanna region of Ghana where bush swidden-fallow farming is practised. The member of the land owning group who is the first to cultivate a piece of unclaimed land establishes for himself and his descendants the right to use that land for cultivation and for those rights also to extend to the swidden-fallow vegetation. Economically useful trees, particularly *Butyrospermum parkii* (sheanut), *Faidherbia albida*, *Adansonia digitata*, *Parkia clappertoniana* (*dawadawa*) and others are integrated with food crops. In areas where there is population pressure on land only these economic trees are left, and most of the trees are now very old.

15. Bertrand, A., 1985. 'Les nouvelles politiques de foresterie en milieu rural au Sahel. Réglementations foncières et forestières et gestion des ressources ligneuses naturelles dans les pays de la zone soudano-sahélienne.' ['New forestry policies in the rural Sahel. the legal status of land and forests and the management of woody vegetation in the countries of the Sudano-Sahelian zone] *Revue Bois et Forets des Tropiques*, no.207. 1er trimestre, pps. 23-40.

REGIONAL The francophone Sahel is addressed in this article. The main economic activities are livestock keeping and agriculture, the balance depending on rainfall. The author describes the way that originally, sahelian inhabitants had their permanently settled villages, their fallows and nearby village owned woodlands, and the open bush between villages. In 1906, the countries of the francophone Sahel came under the French Civil code which stated that all assets which were 'unoccupied and without owner' belonged to the State. Thus all land except permanently settled village land passed out of customary tenure arrangements. A further series of addenda attempted to consolidate this position, and all customary tenure was automatically rescinded in 1956. In no francophone Sahelian countries except Senegal (which had a land reform programme in 1964), have there been changes in forest law since the end of the colonial period. The forest is seen primarily as the place for pasturing animals. There is also heavy demand for fuelwood and construction wood. By contrast with European notions of forests as dedicated pieces of land managed for wood, Sahelian notions are of areas in which other human activities take place in symbiosis with the forest. The sole exception is 'sacred woods' which are strictly preserved by people.

Population growth in the Sahel is of the order of 2% or less in rural areas and 6-8% in urban areas. Concentrated and rapidly growing urban demand is devastating the rural areas for firewood. Long swidden-fallows are disappearing and, in addition, urban entrepreneurs are keeping cattle for cash reasons near towns, where before they would have travelled with them over a wide area. The need to preserve forest needs to be approached along with rural problems so as to reestablish the equilibrium of the agro-silvo-pastoral economy, which is the only one capable of responding to long-term needs. The Sahelian inhabitant will respond only if he has some right of access to forest products. It is necessary to recognise too, as became clear in Senegal, that land tenure laws and administrative and territorial administration are inseparable and must be developed together. Customary tenure needed to be recognised again, but so too do the overwhelming present-day forces, above all the problems of urban growth.

The article contains an interesting bibliography.

16. Bertrand, A., 1986. 'Evolution de l'élevage et politique forestière en zone soudanienne. L'exemple de la 3e Région du Mali', [The development of herding and forest policy in the Sudanian zone. An example from the

Third region of Mali]. Communication presentee au seminaire 'Relations Agriculture-Elevage', DSA-CIRAD, Montpellier, 10-13 September 1985. In *Les Cahiers de la Recherche Développement*, N° 9-10, 1986, pp 35-39.
MALI The author describes the Third area of Mali (Sikasso, Koutiala, Kadiolo) with 'normal' yearly rainfall varying between 900 and 1300mm and recent deficits of 200 to 300mm/yr in a region which is climatically favoured. Here, livestock raising is practised by sedentary herders around villages, by nomadic herders and by urban dwellers. Cattle, sheep and goats are herded. Woody vegetation provides complementary fodder and it can be increased in volume by the reduction of herbaceous competition, i.e. by grazing, provided that rainfall is sufficient.

However, groundwater deficits in upland areas where soils are thin have resulted in tree species like the *Butyrospermum parkii* and *Parkia biglobosa* being endangered and over large areas where these deficits are present natural regeneration has not taken place for years. Forest degradation has been caused by livestock raising, bush fires, sedentary agriculture, shifting cultivation, wood harvesting (in areas near to urban centres) and drought conditions.

17. Bird, N.M. and G. Shepherd. 1989 *Charcoal in Somalia: a woodfuel inventory in the Bay region of Somalia*. Final report of the Energy Planning Study, ODA - UK, prepared for the National Range Agency, Ministry of Livestock, Forestry and Range, Mogadishu, Somalia. ODNRI, U.K.
SOMALIA (see [94] for a summary of this documents also).

18. Bonkoungou, E.G. and R. Catinot, 1986. 'Research on and development of natural regeneration techniques for the silvo-pastoral management of existing forest resources', pps. 89-124 in Carlson and Shea, eds. *Increasing the productivity of multi-purpose lands*, papers from a IUFRO Research Planning Workshop for Sahelian and North Sudanian Zones, IUFRO, Vienna.
REGIONAL The Sahelian and north Sudanian zones still possess extensive mixed forest covering perhaps 50 million ha. At 600-800mm: Intensive agriculture (millet,sorghum, peanut) with part transhumant part stationery animal rearing. At 400-600mm: Agriculture in cooler areas, animal rearing, mainly transhumant, dominant. At 200-400mm: Agriculture near water, animal rearing, nomadic or transhumant, is predominant. Wood production is more important in the higher rainfall areas, forage in the lower.

There is very limited experience among foresters of silvopastoral management based on natural regeneration techniques for mixed forest and grassland. The authors flag the importance of collecting indigenous knowledge in this area, but have none to relay. Research is needed on the role of fire in bushland management, the role of pasture systems in bushland management, ways of speeding the fallowing process through agroforestry. The State is badly constituted for these areas: its interests are too sectoral.

19. Boudet, G., 1979. 'Quelques observations sur les fluctuations du couvert végétal sahélien au Gourma malien et leurs conséquences pour une stratégie de gestion sylvo-pastorale'. [Some observations on variability in Sahelian vegetative cover in Gourma, Mali, and its consequences for silvopastoral management] *Revue Bois et Forets des Tropiques* no 184, March-April 1979, pps.31-45.
MALI The paper discusses the Gourma region of Mali, with a yearly rainfall varying between 128 and 373mm in Gao and between 182 and 345mm in Gossi. Erosion and desertification have resulted in the disappearance of *Acacia nilotica* along rivers and of thickets of *Acacia laeta, Acacia seyal* and *Grewia villosa* in swamp forests. Also, over rocky terrain, during a period of 8 years, there has been continuous forest degradation in upland areas but woody regeneration on the lowland areas. On sandy soils, drought conditions have resulted in openings in the forest cover with *Acacia tortilis* taking over *Acacia laeta* and *Acacia senegal* along gently sloping areas.

20. Breemer, J.P.M. van den, 1989. 'Farmers' Perception of Society and Environment, and their Land Use. The Case of the Aouan in Ivory Coast', *BOS Nieuwsletter*, Vol. 8, N° 18, pps. 28-44.
CÔTE D'IVOIRE This paper discusses a forested region in the basin of the Comoe river. The Aouan are cultivators of maize, banana, manioc, cocoa and coffee. They are sedentary and village-based, practising swidden-fallowing. In former days no importance was attached to personal land rights. Individual land claims arose when cocoa and coffee cultivation, requiring large fixed plots of land farmed individually, gained importance. A demise in agricultural cooperation coincides with the increasing individualisation of tenure.

'Bo' (the forest) provides humans with food and drink, medicinal plants, construction materials, tools and formerly clothes. The Aouan make a distinction between forest and village, distinguishing therefore plant, animal, and human life. Only original residents may lay out new fields in the forest: this is a right connected to matrilineage and the goddess Assie. Rules and taboos prevail concerning the crossing of the boundary between village and forest, and also the laying out of fields: eg fields may not be established on steeply-sloping lands (prone to erosion) because Assie appears there, and this is dangerous to humans. A system of prescribed rest days also operates, when humans may not enter the forest, which is seen as also needing a rest. Bans on certain plants and animals (eg rice and goats) which can be regarded as environmentally destructive also pertain. Rice requires a great deal of sunlight, and it entails widespread clearing of forest areas; as a result, the sunrays reach the soil directly, causing dehydration and erosion. Goats also cause erosion, and destroy vegetation.

In the late 1960's, rainfed rice cultivation was introduced, devastating huge areas of forest. Breemer suggests that conversion to Islam and Christianity, and hence a diminution of regard for indigenous rituals and taboos on forest and land use and the authority of elders, is partially responsible for the adoption of what he sees as a maladaptive ecological practice. Parts of the forest, previously prohibited, became available for agricultural exploitation and perennial cultures were introduced. Hierarchical patron - client relations were undermined when foreign village dwellers were permitted by the government to plant rice, receiving cash and power and escaping from dependence on the indigenous Aouan matrilineage.

21. Brokensha, D., 1986. m.s. *Local Management Systems and Sustainability*. Paper prepared for the Annual Meeting of the Society for Economic Anthropology, Riverside, California. Held by the Institute for Development Anthropology, University of Santa Barbara, California.
REGIONAL The writer focuses on Africa south of the Sahara, and on shifting cultivators. He asserts that in indigenous resource management, there is seldom a formal system or strategy. Instead, there are strategies controlling access to resources which have management consequences dependant largely on the relation of population size to carrying capacity in the existing modes of production. If the former is smaller than the latter, than there is more room to manoeuvre. Brokensha suggests that indigenous resource management systems were never fixed. Instead, he sees the farmer as an innovator when innovation was called for. The success of indigenous resource management strategies inheres in the calculation of the advantages and disadvantages of risk taking or resistance to change. Local agricultural systems are flexible and adaptive.

22. Brokensha, D. and Castro, A.H.P. 1987. *Common property resources*. Background paper presented February 1988, Bangalore for the Expert Consultation on Forestry and Food Production/Security, FAO, Rome. 31pp
KENYA The authors discuss Kirinyaga, Southern Mount Kenya, extending down to the semi-arid Mwea plains. Here, the Mbeere used forests for building materials, medicines and honey, groves used as

places of worship and sacrifice. The sacred groves 'matiiri' essentially belonged to the people of the district, though under the control of the ruling generation set, and they acted as focal point for the concerns of the local community. In the 1930s there were 200 sacred groves within the district, mainly on hilltops or along ridges. With an average size of 0.2-3.0 acres, they were composed of large spreading trees with an understorey. Major sacred species were *Ficus natalensis*; to a lesser extent *Ficus waekfieldii* in the moister areas and *Ficus capensis* in the drier southern lowlands. Except for taking cuttings to propagate new sacred trees, no cutting, clearing or cultivation was allowed in the groves. Calamities were often attributed to illicit cutting. If caught violating the rules, the offender had to pay a goat to the elders and face the wrath of neighbours.

The cultural significance of the groves eroded by weakening of the generation-set system, new religious practices, disappearance of communal celebrations and increased privatisation of land tenure. In the 1930's the groves were no longer sufficiently protected by custom and peer pressure, and they came under the protection of the Local Native Council, administered by the Colonial Government. However the trees have continued to be protected by community sentiment but the groves are essentially relics of a past era. The area also contained islands of forest, the largest of which was 18 kms^2. By order of the local elders these areas had been preserved up to the 1930's to provide structural wood. Rightholders were free to limit the felling of trees and permission from clan elders was needed before larger timber trees could be cut. The management system broke down with migration into the area in the later 1930's. Concern from the colonial administration over deforestation lead to Embu council taking control of the area as a public trust land and later managerial control was transferred to the Forest Department.

23. Campbell, B.M., 1986. The importance of wild fruits for peasant households in Zimbabwe. *Food and Nutrition* 12(1) 38-44
ZIMBABWE The author notes the selective maintenance of favourite wild fruit species in Zimbabwe. As a response to declining availability of preferred wild fruit species in the most severely deforested areas, certain species were selectively maintained. Thus the frequency of consumption of the valued species did not depend on the conditions of the forest area. However deforestation did affect the prevalence of use of the less favoured wild fruits. (This article is referred to in Falconer, J., 1987. *Forestry and diets.* Background paper presented February 1988, Bangalore, for Expert Consultation on Forestry and Food Production/Security, FAO, Rome.)

24. Castro, A.H.P. and Brokensha, D. 1987. *Institutions and foodsecurity: implications for forestry development*. Main paper presented February 1988, Bangalore for Expert Consultation on Forestry and Food Production/Security. FAO, Rome.

KENYA The subject of the paper is Mbeere, Kenya, a physically marginal area with uncertain rainfall and generally poor soils. It was initially savanna woodland. Until well into the colonial era, with a low population and an abundance of woodland, rights to woodland were not regulated. Even by 1970 probably 90% of Mbeere material culture came from woody vegetation, shrubs, lianes and grasses. In pre- and early colonial times some trees were recognised as individual property, especially building trees that had been individually planted like *Melia volkensii*. There was evidence of 'inadvertent conservation' for example sacred groves, the pollarding of trees and the careful propagation of desired species. But the degradation of the forest resource increased with rising population, and the introduction of improved communications, which led to production and curing of tobacco and facilitated the sale of charcoal to urban residents.

25. Centre Technique Forestier Tropical, 1988. '*Faidherbia Albida*. (Del) A. Chev. (Synonyme: *Acacia albida*). Monographie', CIRAD, pp. 29-36 & 50-61.

REGIONAL The paper looks at the value of *Faidherbia albida* in East and West Africa, among sedentary cultivators and herders. This tree provides African populations with a range of uses. It is used as fuelwood, fodder and as windbreak. The seeds are edible, the bark is used for saddle making, the wood for hive and hut construction, and the ashes for soap making. Parts of the tree have medicinal value. This tree is also beneficial as a shade tree for livestock and can also increase soil fertility. *Faidherbia albida* is grown in a productive system in association with corn, millet and sorghum, and also as a shade tree for coffee plantations in Tanzania. In the Sudan, Fur cultivators' law system forbids cutting of the species. Also, Sultan Tanimoun in the 1860s in Niger decreed that anyone caught cutting a gao would be beheaded and that anyone found maiming the tree would have an arm cut. This law resulted in high densities of 100 to 120 trees/ha which led to the disappearance if swidden-fallows.

Because such local laws are not in effect any more, degradation within these man-made forests has resulted due to excessive lopping and pruning, and ageing of the trees. Because of the absence of leaves during the rainy season and deep rooting, the presence of the trees does not affect agricultural output negatively. However, upkeep of the crops under *F. albida* becomes more important.

26. Charter, J.A. and Kay, R.W.J. 1960. 'Assessment of the Olokemeji Fire Control Experiment 28 years after institution.' *Nigerian Forestry Information Bulletin* (new series) No. 3. Lagos, Nigeria. Booklet, 32pp.
NIGERIA The experiment reported here was conducted in savanna woodland close to, and derived from, closed forest. After 28 years: in the fire protection plot, savanna grasses had been eliminated, and 46% of tree stems were of fire-sensitive rain-forest species. The changes in the plots which had been early burnt were going in the same direction, but more slowly. The late burnt plots had remained as open savanna woodland. The paper has been included because it provides interesting confirmation of herder rationality in firing rather than protecting savanna woodland, in order to keep it open and with a good grass cover.

27. Chavangi N.A., Engelhard R.J., Jones V. 1985. *Culture as the basis for implementing self-sustaining woodfuel development programmes* The Beijer Institute and the Kenya Woodfuel development Programme.
KENYA The paper is concerned with Kakamega District in the Western Province of Kenya. This is a densely populated area (up to 1000 inhabitants/sq km) with average farm size of less than one hectare. Most of the land is under private ownership with little remaining communal land. Inhabitants practise agricultural cash-cropping and off-farm labour. Originally land was clan property, 'owned' by the clan head and used by clan members. Today, men own land in their own right. Trees were only ever planted by the male land user/owner, and disputes about land were resolved in favour of the male who could claim ancestral ownership of the most mature trees. Even today, trees are male property because they are linked with the demarcation of farm boundaries and with the provision of the home, which must be exclusively constructed from poles grown by men. Trees are managed by gender access and by species choice.

28. Club du Sahel, 1981. *Analyse du secteur forestier et propositions:le Niger.* [Analysis of the Forestry Sector, and Proposals: Niger] Comité permanent inter-états de lutte contre la sécheresse dans le Sahel. OECD. 2434a Vol. I: Rapport, 2434b Vol. II: Annexes 1-9, 2434c Vol. III: Annexe Juridique.
NIGER Throughout the country 4 climatic areas have been identified with yearly average rainfall varying from less than 200mm to more than 950mm. The natural forest comprises thorn savanna, bush savanna with various *Acacia spp*, tree savanna with *Boscia, Combretum* and *Guiera* and open woodlands with *Tamarindus, Bombax costatum, Parkia biglobosa, Butyrospermum parkii, Parinaria macrophylla* and *Bauhinia spp.* The

inhabitants are sedentary or nomadic herders. Multiple uses for trees include fuelwood for cooking, charcoal for iron work, local timber, utensils, tools, masks, food (fruit, roots, sheanut butter, *Adansonia digitata* leaves), fodder (during the dry season), protection (including windbreaks, live fencing, shade) and other uses such as medicine, ropes, poison, tannin, resin and gum arabic).

In the past, customary rights over forested land resulted in strict land management systems under village chiefs, wise men, and individual families. With colonisation, the system has weakened, and has not been fully replaced by modern ownership laws, so that conflict between government and local villages over land management has resulted.

29. Club du Sahel, 1982a. *Analyse du secteur forestier et propositions: le Mal*i. [Analysis of the Forestry sector and proposals: Mali] Comité permanent inter-états de lutte contre la sécheresse dans le Sahel. OECD. 2435a Rapport, May 1982. 2435b Annexe, May 1982.
MALI The document focuses on Mali with an average rainfall of from 246mm in the north to 1295mm in the south. Herding is the main activity, but 90% of Malians also practice agriculture. High value is given to the fruits of *Parkia biglobosa* and *Butyrospermum parkii,* in the south. Leaves and fruits from *Adansonia digitata* are edible, and its bark is used as rope. *Faidherbia albida* is found where agriculture takes place, and *Tamarindus indica* and mango trees are present around villages. Woodland degradation and desertification are prominent around towns, and erosion and depleted soils in rural agricultural land areas are also environmental problems. Forest stands are over-exploited in many areas in the country. Pasture lands in the Sahelian area are destroyed by excessive pruning and lopping of trees and degraded by bush fires. However, inaccessible forested areas are still protected and the vegetation in the Sahel can be very resistant.

30. Club du Sahel, 1982b. *Politiques forestières au Sahel. Constraintes. Couts*. Organisation. [Forest policy in the Sahel. Constraints. Costs. Organisation] Comité permanent inter-états de lutte contre la sécheresse dans le Sahel. OECD. 2436, November 1982.
REGIONAL The Sahel as a whole is investigated. Originally, Sahelian populations regulated soil utilisation, hunting and fishing, while pasture lands and nomadic herders were regulated by monarchs or other traditions. Historically, certain useful tree species such as *Butyrospermum parkii, Parkia biglobosa* and *Adansonia digitata* were conserved either spontaneously or through strict regulations as is the case for *Faidherbia albida* in the Segou Kingdom (Mali) or in the Sultanate of Zinder (Niger).

During the colonial period, the introduction of crops like the peanut has led to excessive clearing of forests while urbanisation has resulted in degradation around towns. The loosening of ancient traditions during that period also increased environmental degradation.

31. Club du Sahel, 1984. *Analyse du secteur forestier et propositions:la Haute Volta.* [Analysis of the Forestry Sector and proposals: Burkina Faso] Comité permanent inter-états de lutte contre la sécheresse dans le Sahel. OECD. 2436, February 1984.
BURKINA FASO The report deals with Burkina Faso, with its semi-arid climate, long dry season and short rainy season. Ninety-one per cent of the population lives in rural areas from subsistence agriculture and livestock raising (until 1970 the most important export resource). Drought conditions, bush fires and herding have modified the dry closed forests of ancient times into open woodlands. Villages up on the Mossi plateau have been forced very reluctantly to cut *Butyrospermum parkii* and *Parkia biglobosa* trees for fuelwood, once population pressure on the land had shortened or ended swidden-fallow periods in those areas.

In the realm of village forestry it has been found that individual plantations are more popular than community plantations which suggests to the authors that there is greater motivation within the family than within the community.

32. Cordonnier, A., 1988. 'La Foret Villageoise: Modèle de gestion collective des Espaces Boisés du Sahel', [The village forest: a model for collective management of the woodlands of the Sahel] *Annales de Gembloux*, N° 94: 275-287.
MALI The Southern region of Mali (Sikasso). 'With respect to customary rights, forested areas are rights of way which belong to everyone and their ownership to no-one. However, with the introduction of colonialism and the written law, the ownership of these vacant lands was transferred to the State'. Communities, by losing their long-term usufruct to these lands, stopped using them sustainably. Since the 1970s, successive drought conditions have led to degradation of natural forests, and to increased desertification and human suffering.

33. Delwaulle, J.C., 1975. 'Le role du Forestier dans L'Aménagement du Sahel', [The role of the forester in the management of the Sahel] Revue *Bois et Forets des Tropiques*, N° 160, pp 3-22, March-April 1975.
REGIONAL The Sahel is delimited by yearly rainfall of 200 to 500mm; it comprises few natural forest formations except along waterways. Scattered trees are present throughout the savanna, including dense

regeneration of various *Acacia spp.* such as *Acacia tortilis, F. albida, A. ehrenbergiana, A. laeta, A. nilotica, A. sieberiana.*

Up to 1968, livestock raising increased steadily and natural bush regeneration was abundant. Since 1968, however, little regeneration has been possible and woodlands are ageing. *A. tortilis* and *Commiphora africana* have been dying, woodlands have opened up and erosion has set in. By the 70s, gum stands from the 40s and 50s had almost disappeared and no regeneration was present, except for small pockets resulting from the rainy season of 1969. Protection of these stands would be much more economic than new plantations.

34. Dielen H. 1982. *Report of an agroforestry survey in three villages of northern Machakos, Kenya.* Wageningen Agricultural University, The Netherlands and ICRAF, Kenya.

KENYA The report considers the Kamba area of Kenya, with rainfall averaging between 700-900mm. The land has marginal agricultural potential, having a natural vegetation of dry woodland savanna. The major crops are maize and pulses and some animals are kept.

In the past, only farmed land was privately owned, and grazing land was open for all Kamba. Management has moved from the clan to the household, or in the case of shared grazing areas, by particular limited arrangements between a small number of households. With the change from agro-pastoralism to subsistence farming has come a change to privately owned grazing areas which are also a source of firewood and other tree/shrub produce. Several households will share the privately owned grazing areas but rights in trees in the area remain individual.

35. Duijl, Erika van 'Women's access to trees and women planting trees: A study on the complications of tenure issues in agroforestry' in Lucienne Berenschot (ed.) *People's participation in social forestry projects* Bos-Document 7, Department of Forest management, Wageningen Agricultural University and Foundation for Dutch Forestry Development Co-operation. Report of a workshop held on 27 October 1987.

KENYA The study was made in Machakos District among the waKamba, at an altitude of 1200m. This is a sub-humid to semi-arid area with 850mm rainfall per annum. Kamba men used to be hunters, livestock keepers and long-distance traders, while women farmed millet. They have turned increasingly to sedentary agriculture as population densities have increased and land has been registered. The ownership of trees coincides with the ownership of land, and the head of household controls both. The household head's position has become more important with the privatisation and registration of land. There is now evidence that the division of land between sons occurs earlier than previously to encourage greater investment in and improvement to land such as terracing, and tree-planting. Exotics, and planted trees in general, are regarded as private property, while indigenous naturally occurring trees are still regarded by some as available for fuelwood to all. In cases where trees are scarce, the male head of household will control the rate of cutting. Tree-lopping for fodder and for fencing (carried out by men) takes priority over lopping for fuelwood (done by women). Trees may be retained by the individual most easily if they are planted or exotic. Valuable exotics such as *Grevillea robusta* and *Eucalyptus spp.* will be lopped in such a way that a central pole for construction timber is encouraged, while side branches can be used for fuelwood.

36. Ellis, J.E., Coppock, D.L., McCabe, J.T., Galvin, K. and Wienpahl, J. 1984. Aspects of energy consumption in a pastoral ecosystem: wood use by the South Turkana. *In*: Barnes, C. *et al* (eds.) *Wood, energy and households: perspectives on rural Kenya.* The Beijer Institute and the Scandinavian Institute of African Studies , Sweden. 164-187.

KENYA In this area there is a mean annual rainfall of 283-394mm with a 9-10 month dry season. A mixture of vegetation types include heavy bush of *Acacia reficiens* and other *Acacia spp.* with open grass areas, large trees of *A. tortilis* and *A. elatior* in riverine forests, and mountain areas with greater precipitation with a dense shrub forest dominated by *Commiphora* and *A. tortilis.* The report suggests a relative abundance of wood resources for the Turkana with no evidence of previous negative impacts of these people on tree production or density. The Turkana are pastoralists with 2/3 of livestock biomass as browsers (camels and goats) and 1/3 browsers (cattle, donkeys and sheep). They are very mobile and move according to the seasons. All plant materials, including trees are the common property of all individuals in sub-section 'territories'. Women are responsible for the collection of fuelwood and wood for construction. There are few constraints on the collection of wood as this is a common resource. However tree conservation is practised; no live trees of any species are cut for fuel, only dead wood is collected and although live trees are cut for construction, these are usually small trees of abundant species. Although some trees are completely utilised or killed in obtaining construction materials, others are only pruned. The most frequent species used for the construction of dwellings and corrals are *Acacia tortilis* and *A. reficiens* - both are very abundant species in the ecosystem and evidence suggests that they are not decreasing in South Turkana. *A. reficiens* is used in over 80% of huts and corrals. This regional selective cutting may be acting as a partial control on the proliferation of a species that may form dense bush not well suited for livestock when it is not cut, burnt, grazed or otherwise removed.

37. Evans-Pritchard, E.E. 1974. ed. *Man and Woman among the Azande.* London:Faber and Faber; 1974. (Note: based on field work between 1927-1930)
SUDAN The Azande live in the southern Sudan, in the sub-humid zone. They are shifting cultivators with few animals. Women do most of the cultivation of eleusine, yams, pumpkins, gourds, groundnuts and maize. Men plant bananas and fig trees for bark cloth. Trees are needed for fuelwood, house construction, barkcloth (at that date) and for the raw materials for tools and domestic utensils. Land ownership is created by the clearing of land and this is always done by men (since the act creates ownership). Similarly it is men who build huts. Men make barkcloth and have special responsibility for the propagation of the *Ficus natalensis* trees from which it is made.

38. FAO, 1981. *Agroforesterie Africaine*, Une étude préparée par la Faculté des sciences agronomiques de l'Etat, Section foresterie des pays chauds, Gembloux, Belgique. [African agroforestry. A study carried out by the faculty of State agronomic science, Tropical Forestry section, Gembloux, Belgium.] FAO/SIDA programme forestier pour le développement des collectivités locales - GCP/INT/347/SWE.
REGIONAL This study focuses on West and Central Africa. It offers the following examples of indigenous agricultural systems involving trees:
Zaire : At Kimban, in the Kalahari savannas, bamboo is used as live fencing and as fire breaks.
Burkina Faso : for the Mossi farmer, the tree is part of the agricultural system. It provides organic matter, fruits (*Parkia biglobosa, Tamarindus indica, Butyrospermum parkii*), wood and shade. The Dagari farmer maintains forest parks of *Parkia biglobosa* and *Butyrospermum parkii* comprising 30 to 35 tree/ha. These forest parks disappear in proximity to villages and *Adansonia digitata, Ficus spp.* and *Faidherbia albida* are then found.
Senegal: Serer farmers utilise forest parks of *F. albida* for shade, fodder, and for soil enrichment which they provide directly and indirectly. In proximity to Wolof villages in the Bas-Ferlo region, certain trees are protected (such as *Balanites aegyptiaca* from which oil and fruit are consumed and branches and roots used in construction).
Dahomey : Cattle are tied to *Elaeis* palms to manure them. *Butyrospermum parkii* are also grown in the western savannas.

39. Francis, P. 1987. 'Land tenure systems and agricultural innovation: The case of alley farming in Nigeria' pps. 305-319 in *Land Use Policy*, July 1987.

NIGERIA The paper takes the case of the Okwe area of Imo state, Nigeria. If trees are deliberately planted, they belong to the person who planted them. Self-sown trees on individual holdings are the exclusive property of the landholder. Economic trees growing wild on communal land belong to the land-holding group as a whole. An individual farming an area on which tree is growing, has exclusive rights over the tree during period of cultivation.

There are three main categories of land:

- Around the compound: there is a multi-storey cropping system with kola, pawpaw, coconut, and citrus trees over annual crops. The compound area is intensely cultivated with household waste and animal manure.
- Near farms: here cassava, *Telfaria* and other vegetables are cultivated in alternate years. Some of the land may have palms and fruit trees.
- Distant land: grows cassava, maize, melon and bitter yam. There is a six year cropping cycle. Here too stands of raffia palms are preserved.

Land ownership is predominantly individual but there are collective mechanisms for the management of the six year rotations on the distant land by adjacent subvillages.

40. Gibson D.C. and Muller E.V. 1987. *Diagnostic Surveys and management information systems in agroforestry implementation: A case study from Rwanda.* Working Paper no.49, ICRAF and CARE, 1987.

RWANDA This is a study of the transition zone between the hilly Eastern Plateau and the Eastern Plains of Rwanda. The altitude is around 1500m. and the annual rainfall about 1000mm. Farmers live on well-dispersed homesteads. Major crops are beans, corn, sorghum, sweet potatoes and bananas. Livestock are also important, and cattle, goats and sheep are kept. Fuelwood is the main source of household energy, used for brewing sorghum beer, space heating, lighting and ironing. Trees and shrubs provide the raw materials for construction, tools, enclosures and windbreaks.

The 1962 constitution and legislation states that all land in Rwanda belongs to the State. Individuals have exclusive usufruct rights, not outright ownership. Where there is good access to forest resources, there is little tree-planting, and apparently little attempt at management. One of the factors determining tree-planting rates is population density. It is most found where there is the most pressure on land and the greatest intensification of cash-oriented agriculture. Increasing shortages are causing men to take a greater interest in the production of trees. The surveys carried out showed that those male farmers who get fuelwood from their farms have planted more trees than those who collect dead wood. Those farmers who have to cut trees for fuelwood themselves, are more interested in planting trees than farmers whose wives collect fuelwood.

41. Gluckman, M. 1951. The Lozi of Barotseland in north-western Rhodesia. *In*: Colson, E. and Gluckman, M. (eds.), *Seven Tribes of British Central Africa*. Oxford University Press, London. 1-93.
ZAMBIA The Lozi live on a largely treeless plain. They are sedentary farmers and fishermen with an elaborate political structure. Each Lozi king would select or build a village where he would be buried; upon his death, his subjects would erect a fence of pointed stakes around the site and plant trees from the bush there. Thus the grave-sites are clearly recognisable in the largely treeless plain. (Source identified from Niamir, M., 1989. Local knowledge and systems of natural resource management in arid and semi-arid Africa. SIDA/FAO Forest, Trees and People Program.)

42. Grandin, B.E. 1987. East African pastoral land tenure: some reflections from Maasailand. *In*: Raintree, J.B. (ed.), *Land, trees and tenure*. ICRAF and Land Tenure Centre, Nairobi and Madison. 201-210
KENYA The paper focuses on Maasailand, East Africa, and an aspect of the system before group ranches were established. Trees on calf pastures, land that was set aside near residential sites which provided grazing for young calves or sick or old animals, and have some special rights attached to them. Permission was not required for the cutting of dead trees for firewood, or to cut a few living branches (e.g. for a walking stick), or to take a few roots for herbs. The only time permission of the controller(s) of the calf pasture would need to be sought was when massive amounts of branches for fencing was to be cut.

43. Griffard, P.L. 1964. Les possibilities de reboisement en *Faidherbiaalbida* au Senegal. [The possibility of reafforestation with *Faidherbia albida* in Senegal] *Bois et Forets des Tropiques*. 95, 21-33.
NIGER The Hausa sultans of Zinder in Niger set themselves up as protectors of the 'Gao' (Faidherbia albida) decreeing draconian preservation measures. Anyone cutting down a tree without permission had his head cut off, and anyone who mutilated a tree without good reason would have his arm amputated. There is also mention of contemporary use of *Faidherbia albida* in an agropastoral system with 10

to 50 trees per hectare. (This article is cited in: Boudet, G.C. and Toutain, B. 1980. The integration of browse plants within pastoral and agropastoral systems in Africa. In, Houérou, H.N. le (ed.), Browse in Africa: the current state of knowledge. International Livestock Centre for Africa, Addis Ababa. 427-432.)

44. Gulliver, P. H. 1958. *Land tenure and social change among the Nyakyusa.* East African Studies no.11, King and Jarrett, London.
TANZANIA An area located north of Lake Nyasa, in Rungwe District, Tanzania. The altitude ranges from 1550-8000 feet, with good rainfall. The main crops were bananas, maize, beans, finger millet and groundnuts. The highest areas have planted coffee as a cash crop, while down in the lake plains paddy rice was both a food and a cash crop. Livestock were also important. The village was the basic land holding unit, under a headman. Residency gave rights to a house-site, banana plot, arable land and grazing on communal pastures. A man who left must give up these rights but could re-activate them later. Management was by the exclusion of non-residents - even if they were from the same chiefdom or clan. When a chief died trees were planted on his grave, forming a sacred grove and religious centre where later sacrifice might take place.

45. Hammer, T., 1982. Reforestation and community development in the Sudan. *Energy in Developing Countries Series* Discussion Paper D-73M, unpublished, Resources for the Future, Washington.
SUDAN The paper describes the Kordofan region of Sudan, with 200-800mm rainfall annually. This was once an area of exceptionally high fertility, but soil fertility began to decrease 30-40 years ago. Important local tree species include: *Acacia senegal, A. tortilis, A. mellifera, A. tortilis, F. albida, A. seyal, Combretum cordofanum, Zizyphus spina-christi* and *Balanites aegyptiaca*. The population are sedentary farmers with animals. Originally, shifting cultivation was practised, with 3 plots on a rotational basis. A plot was cultivated for 3 to 6 years, with grazing of the fields after harvest by the cultivators animals or those of nomadic herders for fees. *A. senegal* naturally regenerated and was either left to grow alongside the agricultural crops or cleared until the field was left fallow (very occasionally seeds of *A. senegal* was sown). During the fallow the trees were tapped for gum. In the late 16th century a quasi-privatisation of land developed, in which farmers began to claim rights to plots after the cultivation period in order to secure income from the *A. senegal.*

Until the mid 1950s dead and fallen branches were used for fuel, and living trees were only cut for land clearance. Gum bearing *A. senegal* was never cut. However there has been a general decline in the land's capacity to support the local population since the middle of this century for a number of interrelated reasons desert encroachment from the north over-exploitation of the woodland savanna: overly intensive agriculture, overcutting of wood for fuel and overgrazing. The intensification of pressure of land was caused by increasing indigenous human and animal populations, and migration from the degraded area to the cultivable village land in the core of the Gum Belt. Pressure for land has also made sheikhs less willing to give village land to newcomers. They may be rented land but have no rights to the land or the trees that grow on it after the cultivation period. The tenants therefore have little incentive to try to maintain or improve soil fertility. Due to falling gum yield and increasing local demands for fuelwood live trees including *A. senegal* are beginning to be cut, although sheikhs try to punish those that do. Also there is now demand for cutting of trees for charcoal for markets up to 300km away.

46. Hammer, T., 1988. 'Wood for Fuel - Energy Crisis Implying Desertification: The case of Bara, Sudan', pps. 176-181 in *Whose Trees? Proprietary Dimensions in Forestry*, Fortmann & Bruce (eds), US: Westview Press.

SUDAN The study concerns Bara, in Kordofan, Sudan. Here, *Acacia senegal* constitutes the main tree cover. The inhabitants are sedentary cultivators based in village settlements. The general land distribution practice sees the allocation of three to four plots of land per household, with the reserve remaining under the village head's control, to be allocated to new arrivals, or otherwise designated as common property. Family plots pass from father to son and deserted plots revert to common property.

In practice, village heads are now reluctant to give newcomers plots. When they do so, they do not give the newcomers rights over the gum trees present on their plots, but appropriate the profit from the trees for themselves. They also lay claim to gum trees on open or government land. Each long-established household retains the right to tap its own gum trees, and to prosecute trespassers.

Originally, the cultivation-fallow-gum tapping-charcoal burning system was strictly enforced, with long swidden-fallow periods maintained.

Forest laws, implemented by the Forest Office in Bara, dictated that no trees should be cut except for the purpose of cultivation and with the Office's permission. However, many trees are now cut daily on forest land, without permission. The complete cycle has halved from around 18-20 years to only 8-10, and the swidden-fallow period has contracted from 10-15, to 1-3 years. The shortened swidden-fallow period is both a response to and a cause of worsening ecological conditions. It interacts with a complicated set of factors, resulting from both climatic and human conditions. Diminished rainfall (1968-76) caused a decline in gum yields, so cultivators cut down the trees and began growing crops early in order to make up for lost income.

47. Heermans, J.G. 1988. 'The Guesselbodi experiment:bushland management in Niger' pps. 84-87 in *The Greening of Aid: Sustainable Livelihoods in Practice* eds. C. Conroy and M. Litvinoff. (The proceedings of an IIED Sustainable Development Conference). Earthscan publications, IIED, 1988.
NIGER Guesselbodi forest, 25 km from Niamey. Dominant tree species: *Combretum nigricans, C. micranthum* and *Guiera senegalensis. Boscia senegalensis* and some *Commiphora africana*. In the past, there was also plenty of *Prosopis africana* and *Sclerocarya birrea*. The inhabitants practise a good deal of herding and some agriculture.

Ligneous species in the Sahel sprout readily if lopped, and grow rapidly. Some *Combretum nigricans* cut to ground level, grew three metres over the following year. No other information related to indigenous management.

The Guesselbodi management plan was presented to the people as a fait accompli, though it was based on two years of earlier questionnaires and discussions. It is hard to know how wide-ranging these were. While it is clear that people will not participate in management programs until ownership, control and land tenure questions are resolved, it would seem that here, local people opted for the income generating opportunities the project presented. (i.e. it was not sustainable without these inputs). Management is to be by co-operatives, and responsibility will be 'gradually transferred' to them. The co-op will sell fixed numbers of firewood cutting and grazing permits and this will be used to raise cash for guards' salaries and so on.

48. Horowitz, M. and Badi, K. 1981. *Sudan: introduction of forestry in grazing systems.* FAO/SIDA forestry for local community development programme - GCP/INT/347/SWE. FAO, Rome.
SUDAN The document deals with the sedentary farming communities in the irrigated agricultural schemes adjacent to the White Nile, and semi-nomadic pastoral-cultivators in the sandy uplands to the west. Average annual rainfall ranges from 150mm to 400mm within the region. Generally the woody resource is degraded from the original vegetation types. In the Acacia tortilis -*Maeru crassifolia* desert scrub *A. tortilis* still dominates, but many other species are no longer found. *A.*

tortilis owes its dominance to the firm protection it enjoys from the local inhabitants. The tree is used for browse and fallen leaves and the collected unripe pods are used for cattle, sheep and goat feed. There are also village forest areas close to the village which are closely supervised by the villagers themselves. Wood collection and grazing is limited to actual members of the village. These areas are generally well maintained, except where the villagers of small poor villages have no cash alternative to fuelwood collection. The only privately owned trees result from natural regeneration of trees during the swidden-fallow period. These trees then belong to the title holder of the field.

49. Hoskins, M.W., 1984. 'Observations on Indigenous and Modern Agroforestry Activities in West Africa', pps. 46-50 in *Social, Economic, and Institutional Aspects of Agroforestry*, J.K. Jackson (ed), United Nations University.

REGIONAL The paper deals with sedentary cultivators operating swidden-fallowing systems in West Africa: Niger, Nigeria, Chad, Burkina Faso, and Mali. Hoskins notes the presence of common property resources in several of the countries. Herders and farmers have in the past had complex regulations regarding access to land, water and vegetation. Common lands remain an important source of income for women. Often, planting trees changed land ownership, so tenants might not plant trees on the landlord's property and farmers suspected the motives of forestry departments trying to get them to plant trees. Modern land-use planning programmes tend to privatise land ownership, ignoring the use of natural vegetation by forest-dwellers, farmers, the landless or herders. Titles go to male household heads, ignoring indigenous gender-related usufruct rights. Shea nuts (*Butyrospermum parkii*) are manufactured into cooking oil by women. Leaves and seeds comprise nutritious sauces for grain-based meals; grasses and bark are woven and dyed into mats and baskets, and sold or used at home. Trees serve medicinal functions and provide fuelwood, chew sticks (tooth brushes), ropes and timber. *Faidherbia albida* is important for soil fertility and fodder. Taboos were often placed on certain species, activities, or the use of certain tools for cultivation or collection. In Sierra Leone, swidden agriculturalists cut trees in field clearing operations at various heights from the ground, to favour regrowth of selected species when the fields are once more fallow. In Senegal, Burkina Faso, Chad and Niger *Faidherbia albida* is selectively preserved for its benefits to the surrounding crop. Likewise, *Butyrospermum parkii, Parkia biglobosa* and *Adansonia digitata* trees are preserved in the fields.

Many new agroforestry projects have been implemented in West

Africa. It is suggested that their success rate corresponds to the degree to which indigenous resource management practices are incorporated into the systems imposed by extension agents. In practice indigenous practices and tenure rules are often overridden, so that the new project fails to benefit more than a selected few in the village.

50. Houérou, H.N. le, 1980a. The role of browse in the Sahelian and Sudanian zones. *In*, Houérou, H.N. le (ed.), *Browse in Africa: the current state of knowledge*. International Livestock Centre for Africa, Addis Ababa. 83-100
REGIONAL This review mentions that 'the agropastoral systems developed by several African peasant societies based on..(*Faidherbia albida*)..exploitation and millet and sorghum crops, seem to have arisen independently: for example, the Galla and Arussi farmers in the Ethiopian Rift Valley have developed a park like landscape similar to that of the Serer in Senegal.'

51. Houérou, H.N. le, 1980b. Agroforestry techniques for the conservation and improvement of soil fertility in arid and semi-arid zones. In, Houérou, H.N. le (ed.), *Browse in Africa: the current state of knowledge.* International Livestock Centre for Africa, Addis Ababa. 433-435
REGIONAL *Faidherbia albida* is protected and kept in millet fields in a number of peasant civilisations in Africa, among them the Serer of Senegal. The density of trees is 10 to 50 per hectare, the adult trees are 10 to 20m tall and the canopies cover about 2% to 40% of the ground area. Millet production in the system is almost twice that obtained in purely open-land farming systems without fertilisation. Also there is considerable value of the pods as animal feed. Branches and foliage are lopped on a rotation system about once every 3 to 5 years and fed to livestock. The wood from the branches are used for fuel and fences. The total economic output of the system for the farmer is probably about three times greater than the open-land millet cultivation. In addition, soil fertility is maintained and soil erosion is kept to a minimum, unlike in open field conditions.

52. Huntingford, G. W. B. 1955. *The Galla of Ethiopia. The Kingdoms of Kafa and Janjero.* Ethnographic Survey of Africa, D. Forde (ed), North Eastern Africa, part 2, International African Institute, London.
ETHIOPIA The Galla live on the Ethiopian plateau, and enjoy heavy seasonal rain. They are primarily growers of cereals and legumes. For the Galla, land in theory belongs to the tribe. However, some degree of rights to land and forest can be obtained through occupation 'accompanied by the hanging of honey-barrels in the forest; but no man

may enclose for himself more than he can cover with a spear-throw or stone's throw'. The Galla have a great reverence for Nature and 'groves of trees, and individual trees, especially *Ficus sycamorus*, near huts and villages are indeed sacred'. These are often planted for spirits as an act of piety.

53. ILCA, 1979. Livestock production in the subhumid zone of west Africa: a regional review. *ILCA Systems Study* 2. International Livestock Centre for Africa, Ethiopia.
REGIONAL The study concerns the subhumid zone of west Africa, which has a mean annual rainfall of 1000-1500mm and a growing season of 180 days in the north to around 280 days in the south. Woody vegetation ranges from semi-deciduous forest along streams in the savanna zone to open tree savanna dominated largely by isolated trees. Browse is not considered as important a component of animal diets in the subhumid zone as in the Sudanian and Sahelian zones because of the number of woody species acceptable to stock is limited and grass is generally available. Browse behaviour is very seasonal and is associated with the new flush of foliage which appears on the browse plants with the onset of the hot and humid weather preceding the rains, a period when grass tends to be in short supply. In many parts of the northern subhumid zone herdsmen climb large trees and lop branches for fodder. This supplements other more accessible browse.

54. Jackson, J.K. 1983. *Management of the Natural Forest in the Sahel Region* Technical Report prepared for USDA Forestry Support Program (Technical Assistance Program for AID's Forestry Development Activities) AID/USDA/USFS P.O. 40-319R-3-00273, 02/09/83
REGIONAL The productivity of natural forest is about 0.5 cu.m /ha/yr in the Sahel zone rising to about 1 cu.m /ha/yr in the Sudan zone. Tree cover ranges from : *Acacia tortilis* in the driest areas; in slightly wetter areas: *A. senegal*, *A. laeta*, *Commiphora africana*, *Ziziphus mauritiana* and *A. seyal* on heavier soils; under higher rainfall still: *Combretum glutinosum*, *C. micranthum*, *Anogeissus leiocarpus*, and *Guiera senegalensis*. Much of this area shows a forest strongly influenced by man, with selected species retained and others cleared. In drier areas these anthropogenic forests contain *Adansonia digitata* and *Faidherbia albida*, while in higher rainfall areas are found *Butyrospermum parkii* (*karité*), *Parkia biglobosa* (*néré*) and *Tamarindus indica*.

Forests are an important reserve of relishes, dietary supplements and fodder, at certain times of year and in times of drought. Naturally they are also important for fuel and poles, fibres and medicines. The people are herders and, where possible, agriculturalists, farming on a long swidden-fallow basis or intensively by rivers and permanent water sources. The legal position as regards forest for rural people in francophone Africa is difficult, because all customary rights are essentially denied and the population feels that the *forets classées* have been taken from them by foresters. There has been little forester management of any part of the Sahel, except for protection and early burning. The main exception is that of *Acacia nilotica* along the Blue Nile in the Sudan. State protection is in its infancy.

Indigenous management in this document, is advocated as a solution to the lack of forestry staff. The report contains an excellent bibliography.

55. Jackson, J K n.d (post 1984) *The management of dry natural forest in West Africa and the Sahel, with some general thoughts on management of indigenous forest*.m.s.

REGIONAL Up to 1983, the author could find only 2 instances recorded of classical planned natural woodland management in Africa.

- *Bandin forest, Senegal.* 10,400ha managed from 1953 for fuel and sawlogs. The plan was for coppicing in coupes over a twenty year rotation. 'The production of sawlogs from *Khaya senegalensis* was to have been based on an annual inventory by sampling, but in fact many of the trees intended for timber production were cut by charcoal burners.'
- *The management of Acacia nilotica on the Blue Nile* There has been planned management since the 1930s for railway sleepers and firewood. The area is annually inundated and seeds germinate readily, helped along by additional sowing.Formal forest management inputs should always be kept low because values are low, but the author suggests some possible approaches:
- Yields can be increased by fire protection, though it may engender conflict with local people. In other circumstances (little research has been done on this) grazing may have protect trees by removing inflammable grasses. Various experiments seem to suggest that occasional (but not annual) early burning, coupled with fire protection, produce the greatest MAI.
- For fuelwood production, simple coppice regulated by area is the most likely to succeed. If local people are to harvest, stem sizes should be kept small to accord with local implements. Probably short coppice rotations produce higher yields than long ones.
- Probably management without a preliminary inventory can be begun if management is for fuelwood, and all that is cut will be used. The exact amount produced will be known slowly. A knowledge of growth rates is far more important than a knowledge

of standing volume.

- Coppice with standards is a possibility for a mix of fuel and larger timber.
- So-called selection systems allow the taking of trees above a certain girth. But so little is known about growth rates in African woodlands, and thus about the proportion of younger trees which should be left to replace those removed, that this selection of larger trees may be destroying the more genetically vigorous individuals.
- The management by local people of forests makes sense because they know best what products they want from the forest and are eager to control them. Pre-requisites for such local management are good relations between villagers and local authorities; and a good understanding of fire and cutting rules.

56. Jean, B., 1985. 'Sécheresse et désertification au Mali. 2e partie: Perspectives', [Drought and desertification in Mali. Part 2: perspectives] *Revue Forestière Française*, N° 37(4): 315-331, Chronique Internationale.
MALI The paper focuses on Mali as a whole. Rainfall deficits of 100 to 200mm (average) and decreases in the water level of the Niger river of 50 to 60 cm have led to lower soil fertility, drying up of the forest cover, modification of the flora in the pasture lands, disappearance of wildlife species, and erosion. Village forestry includes 250 to 300ha of fuelwood plantations. The objectives are to have individual families planting 10 trees per year around the houses.

57. Jelenic, N.E. and Vegten, J.A. van, 1981. A pain in the neck: the firewood situation in south-western Kgatleng, Botswana. *National Institute of Development and Cultural Research (NIR) Research Notes* No. 5. University College of Botswana, Gaborone.
BOTSWANA The paper is concerned with South-western Kgatleng, Botswana. Vegetation is a bushland-shrubland mosaic dominated by *Acacia* species and grass - a vegetation is derived from woodland dominated by *Combretum* and *Digitaria* by the burning, ploughing and grazing activities of man. Soil erosion is common, trees are often stunted and bush encroachment is serious. There is substantial labour migration, mainly of men, to Gaborone and South Africa. Most people live off communal land where they carry out mixed subsistence farming with crops and domestic animals.

The collection of domestic wood is carried out by women. The location of fuelwood collection is primarily governed by the sector of the village that the women live in: those from the east side of the village go only east to find wood. Children and old or sick women collect wood from nearer the village. Only dead wood is collected, and only from selected species, though these are becoming increasingly scarce. To discourage rule-breaking within the groups that collect firewood someone among those who go is selected for their law abiding qualities. Indigenous beliefs protect fruit trees including *Boscia* species, which also produce shade and browse.

Between the two villages in the study firewood supplies are much depleted and people now collect wood from outside their original sectors. Firewood collection rules are now not always adhered to by young people, and during the last ten years the distances walked to collect wood has doubled or tripled. The relatively high removal pressure induces a succession from *Combretum spp.* to thorn bush to total denudation. There is also commercial firewood collection on this communal land. *Boscia* species are used for fuelwood since they are now the only trees left at short distances from the village. The resource is becoming depleted because of overgrazing and the cutting of live trees for building materials and fuel.

58. Kenyatta, Jomo. 1965 (reprinted). *Facing Mount Kenya.* New York:Random House.
KENYA The Kikuyu live on the east and south of Mount Kenya. They are settled agriculturalists and keep livestock. Land is inherited from father to son and every district has communal pastureland for livestock grazing. In addition to communal grazing arrangements, some public land is also preserved as sacred groves. These (often composed of *Ficus natalensis*) are usually on the tops of ridges, and were the venue for meetings of clan elders, and communication with spirits and ancestors in the pre-Christian era. They have in some cases thus become the present-day sites of churches and government buildings. ed.)

59. Kerkhof, P. 1990a. 'Forest Land use project, Guesselbodi, Niger.' pps. 177-186 in *Agroforestry in Africa: a survey of project experience.* (eds. G. Foley and G Barnard) Panos Institute, London.
NIGER Rainfall in this area is about 500mm per year, falling between May and September. The forest and nearby Niamey are on Niger's high plateau. The region of Guesselbodi forest is inhabited by Djerma farmers, and the semi-nomadic Fulani and Tuareg pass through with animals. The forest itself is uninhabited: it is a *foret classée.*

The forest is used for livestock grazing, honey and medicine collection, and contains important long-used livestock routes. It also supplies fuelwood for Niamey. It has declined 50% in 30 years. Only 3,000 of its 5,000ha has any tree cover, and much of that is in poor shape. Various reforestation techniques have been tried, mulching and tillage being the most effective stimulants to improved growth.

Local people were initially uninterested in management and distrustful of the foresters. But their interest grew once the fuelwood co-operative was set up in 1987. It permitted the inhabitants of the nine local villages to harvest fuelwood and sell it to the co-operative, who in turn sold it to Niamey. Villagers could also cut and sell hay. In 1988, the latter earned about US $6,700. The co-operative is still receiving a lot of project help, and may not be financially or socially the best institution for local participation in forest management. Foresters show villagers which trees they may cut a) by marking trees to be preserved b) by handing out diameter-measuring tools which indicate the minimum size which may be cut. However, villagers still see the project as controlling the forest, and their own role as obedience to the rules. Guards (Tuareg) with camel and sword guard 250ha each and animals found in the area are impounded. Such a system has proved much cheaper than fencing, but still costs the project US $130 a month for each guard.

Niamey, only 30km away, has a population growth rate of 10% per year, and currently has a population of 700,000, 90% of whom rely upon fuelwood for biomass. The project is not commercially self-sustaining. Management costs cannot be met from the sale of woodfuel in Niamey, and the prospects for raising that price are negligible. Inessential components of the project such as soil conservation and tree-planting will have to be abandoned, reducing costs merely to guards' salaries in areas where regeneration is feasible. It is not clear whether such projects will ever be able to function without substantial financial inputs from outside.

60. Kerkhof, P. 1990b. Turkana Rural Development Project, Kenya.' pps.161-170 in *Agroforestry in Africa: a survey of project experience.* (eds. G. Foley and G Barnard) Panos Institute, London.

KENYA The Turkana area of Kenya has an annual rainfall of 180-400mm. The dominant species in the riverine forests are *Acacia tortilis* and *A. eliator*, and in the open rangelands, *A. tortilis* again and *Dobera glabra*.

In recent years drought and heavy animal losses have turned the Turkana, to some extent into sedentary agriculturalists on the banks of the river Turkwel. The creation of a tarmac road all the way from Nairobi into Turkana has also changed settlement patterns and exploitation of the environment. Most of the area's rangeland is held communally by the Turkana, especially during the rainy season. During the dry season, much smaller household groups camp with their animals on their home bases, (*Ekwar*) which are segments of the riverine forest running along the banks of the river Turkwel. Apart from household management of their own **Ekwar**, clan elders have always had a general responsibility for the open rangeland and its exploitation. It has been possible for the project to work with elders to ban charcoal export from the district and, in some areas, to manage the resource more elaborately. For instance in the Loregum area, where the *Acacia tortilis* stocks were under intense pressure, new lopping rules restrict cutting to side branches, so that the dominant shoot can grow rapidly above goat browsing height. The result has been 'spectacular regeneration of tens of thousands of young *A. tortilis* trees'.

61. Kessler, J.J. 1990. 'Agroforestry in the Sahel and Sudan zones of West Africa', in *BOS Nieuwsletter* no.20 vol 9(1) 1990, *People and trees in the Sahel*, pps. 27-33.

REGIONAL An article which cautions against too naive an assumption that previous agroforestry systems in the southern Sahelian zones (300-600mm) and the northern Sudanian zones (600-900mm) offer exciting possibilities for the future. Even trees with tap-roots, such as *Faidherbia albida*, still also have numerous upper soil roots which compete with crops and yields are only higher under these trees for the following reasons:

- *A.albida* grows where the water table is not too deep - hence its green leaves in the dry season. It establishes on fertile soils, rather than creating them.
- many tree species have higher soil fertility under them; a wide rooting system brings in nutrients which are dropped under the trees as leaf litter. However, only *A.albida*, because it is leafless in the growing season, can release these nutrients to the plants.
- Because the tree has leaves in the dry season, animals congregate under the trees to browse, drop their dung there - and add to fertility.
- there is no nitrogen fixation: no nodules are formed. So this species does not create additional fertility, it redistributes it within the field.

The farmer will tolerate widely spaced light shade if trees have useful products, but only if he is not too short of land, and only until he begins ox-ploughing and/or cash-crop production.

62. Lai, C. and Khan, A. 1986. 'Mali as a case-study of forest policy in the Sahel: institutional constraints on social forestry'. *ODI Social Forestry Network Papers*, 3e.

MALI encompasses a vegetation range which goes from dense Sudano-Guinean forest in the south, to barren desert in the north.

In tandem with this, Mali's people range from agro-silviculture in the south (trees, cereals and cotton and peanuts as cash crops), through agro-silvo-pastoralism, with sedentary mixed farming systems incorporating crops, trees and livestock, to transhumant herders in the north moving between wet and dry season pasture, and for whom browse forms at least a third of their animals' annual dietary requirements.

Deforestation began to be noticed in the Sahel as early as the 1930s: the result of cash-cropping and wood demand for building and fuelling railroads. As a result, forest policy in francophone West Africa began to be formulated in an attempt to control deforestation by the creation of State forests. The reserves were created in lands regarded as empty, but which in fact were parts of the range lands of pastoralists and the bush swidden-fallows of agriculturalists. Conflict over alienated lands helped to shape the 'punitive and deterrent aspects' of forest law; and was not helped by the fact that a substantial proportion of forestry funds not raised from donors, comes from permits and fines imposed on rural people by forestry staff.

So much has changed in the last sixty years, that the devolution of forest management back again to rural people - unaccompanied, it would seem by the political authority which once went with it -looks fraught with difficulties.

63. Legesse, A. 1984. 'Boran-Gabra pastoralism in historical perspective' in P.J. Joss, P.W.Lynch, and O.B. Williams eds. *Rangelands: a resource under siege. Proceedings, 2nd international Rangeland Congress.* Adelaide, Australia.

KENYA The article describes the Gabra, a cushitic language speaking group living in semi-arid north-eastern Kenya. There are prohibitions and restrictions on certain species such as *Acacia tortilis*. Trees in sacred groves may not be touched at all.

Reference in Niamir, 1990.

64. Malcolm, D. W. 1953. *Sukumaland, an African people and their country: a study of land use in Tanganyika.* International African Institute, Oxford University Press, London.

TANZANIA The Sukuma are settled agro-pastoralists, using grain as currency, and cattle as capital. About half the area is under cultivation and the rest under bushland, in which *Acacia spp.* dominate in the eastern areas, and *Terminalia spp.* in the North West. The rainfall is about 750mm per annum. The chief is the owner of land, but he delegates right allocation to the village headman. Individual rights are limited to the period of effective occupation, and lapse after the harvest of the annual crop; however, planted trees perpetuate the right to a return because of the labour expended.

When a village moves to a new site, the headman gives permission for bush to be cleared, and allocates plots. Honey and beeswax are collected in the bush. The individual lays claim to them by the expenditure of labour e.g. by driving in pegs to climb the tree. Indigenous trees are used for building poles, bark rope, wooden utensils, house-wall uprights, string, baskets, tooth-brushes and hafts for hoes. Access to the resource is limited by the residence of the user.

65. Manoukian, M. 1950. *Akan and Ga-Adangme Peoples of the Gold Coast* Ethnographic Survey of Africa, D. Forde (ed), Western Africa, part 1, International African Institute, Oxford University Press, London.
GHANA The article focuses on some of the sedentary matrilineal agriculturalists of southern Ghana, who derive a cash-crop income from trees. The lineage as a whole owns trees (such as oil palms) which are already growing on family land, with ownership being vested in the head of the lineage. Early in the century, family-owned land used in shifting cultivation was the general rule, with individual ownership very uncommon. However the introduction of tree crops changed this. Land can come to be 'individually owned' in four ways, by a person clearing bush of forest land for the first time, by a man planting economic trees (generally a cocoa farm), by mortgage or purchase, or by inheritance.

66. Marchal, J.Y. 1983. 'Yatenga, nord Haute-Volta: la dynamique d'un espace rural soudano-sahélien', [Yatenga, northern Burkina Faso: the dynamics of a Sudano-Sahelian rural area] *ORSTOM, Travaux et Documents* No. 167, ORSTOM, Paris, pp 170-175.
BURKINA FASO This is a study of Tugu village in Burkina Faso, with its parkland. Mango trees are planted over 1.5ha and are located along the outer edge of the 'parkland' under study., which covers an area of 97 ha. Millet and sorghum are cultivated in association with *Faidherbia albida* (55% of 'park' area), *Balanites aegyptiaca* (15%), *Sclerocarya birrea* (plum tree, 10%), *Tamarindus indica* (4%), *Lannea macroptera* (raisin tree, 3.5%), *A. tortilis* (14.2%), *A. scorpioides* (3%), *Ficus gnaphalocarpa* (fig tree, 14.2%), and 6 other species. These trees originate from natural regeneration and are maintained and protected from grazing even during the dry season. Also, *Balanites* and *Mimosaceae* are removed by local people and replaced with *Butyrospermum parkii* and *Parkia biglobosa* even though these are at their ecological limits. Transhumant Fulbe herders are only allowed to let their animals graze while the farmers who are the tree owners do not hesitate to cut branches for their sheep and goats.

67. National Academy of Sciences 1983. 'Traditional Land Use Systems: energy flow and productivity in traditional sahelian agricultural systems'. Chapter 2, in *Agroforestry in the West African Sahel*, National Research Council. National Academic Press, USA.

REGIONAL The article discusses the Sahelian zone, with 200-750mm rainfall p.a. The people of this area practise either

- pastoral nomadism: with grazing over vast areas, and migration patterns dictated by seasonality; the author argues that these users practise multiple exploitation of plant cover but no cultivation of plants; or
- sedentary rain-fed agriculture, concentrated into a short season dependent on rain. Livestock are kept as insurance. Bush swidden-fallow is the main system of tenure: land is owned by the household or the kin group.

Under silvo-pastoral systems trees provide shade and browse for livestock, (browse is essential both for its volume and its nutritional value), edible fruit and leaves, wood for construction, and medicines.

Under agro-silvo-pastoral systems trees provide fencing and other construction material, fuel and medicines for people and livestock.

Under agro-silvicultural systems trees provide fruit, shade, fencing and fuelwood.

In some silvo-pastoral systems, trees are cut to stump level for various purposes (timber, fuel, browse), then left to recover for several years in a fallow period.

68. Netting, Robert. 1969. 'Marital relations in the Jos Plateau of Nigeria.' *American Anthropologist*; 1969; 71:1037-1045

NIGERIA The Kofyar people live on the edge of the Jos Plateau in northern Nigeria, a high area (4000 ft) of grassland and rocky outcrop. Rainfall averages 1000-1500mm and the vegetation is sub-humid wooded savanna. Intensive hill agriculture on terraces is practised, with shifting agriculture on the plains. The Kofyar keep livestock, and the men hunt. Fuelwood, thatching, poles and raffia leaves are the main woodland produce. Women are forbidden to enter the sacred grove which belongs to each lineage.

69. Niamir Maryam, 1990. *Herders' decision making in natural resource management in arid and semi-arid Africa.* Community Forestry Note 4. FAO-SIDA Forests, Trees and People Programme, FAO, Rome.

REGIONAL Niamir notes that the most common form of tenure in these areas is through the area-management mechanism of the tribe/ clan/ lineage/ household nesting degrees of ownership, relating to progressively more subdivided portions of the range. Trees are vital for forage, building materials, carpentry and carving and, to a lesser extent, fuelwood. She observes that, though there are numerous descriptions of the management of range, there are far fewer published details of the management of the tree itself, or how products are harvested. What she has found, are references to: the policing of how trees are lopped; prohibitions on the use of certain species for certain purposes; prohibitions on use at certain times of year; or in a few cases, ever. She also observes that pastoralists are far more careful of trees on their home range, than they are when passing through the lands of others. Many of her sources note the collapse of pastoral management systems.

70. Norton, A., 1987. *The socio-economic background to community forestry in the Northern Region of Ghana.* ODA Community Forestry Project in the Northern Region of Ghana. m.s.

GHANA The area Norton is describing has a mean annual rainfall of 1000-1400mm a year, and is studded with compound farms and bush swidden-fallows. Ownership of certain species, mainly exotics, is invested in the person who plants, or arranges the planting of the tree (there are local beliefs concerning the 'dangers' of tree planting). The species include kapok, citrus, neem, guava, pawpaw and mango. However if a farmer protects the regeneration of sheanut on his farm this does not give him ownership rights over the trees. Neem is not subject to proprietorial rights in areas where it is common in the bush. It seems likely that proprietorial rights could be established over any species if it were planted and tended by an individual. However this may be difficult with *Parkia clappertonia* due to its strong association with the institution of the chieftaincy. The bush is considered to be a common resource to the members of a local community. Bush burning is sometimes practised, but the reasons for this activity are uncertain.

By custom, only dead wood is gathered for fuel but as the resource is degraded around settlements there is a greater temptation to cut live trees. There is a prohibition throughout the region on the cutting of *Parkia clappertonia* and *Butyrospermum parkii* (Sheanut), though in areas of acute shortage this is ignored. One village had instituted an apparently effective ban on cutting of sheanut for fuelwood and this had led to considerable regeneration of young trees around the village. There is acute shortage of constructional timber in some areas. Apart from fruit trees the growing of neem (*Azadirachta indica*) to supply roofing poles is the only form of tree-planting activity to become widely established in the region over the last 20 years on the initiative of the farmers themselves. Wildings are collected from areas where neem has become established. Rights to natural regeneration of indigenous species depend on the type of area it occurs in. In the 'far bush' land no ownership rights exist at all in trees or their products. In the 'near bush', within the territory of a chief or community, the chief has the right to harvest all the highly valued *P. clappertonia* trees. This is the only form of individual right in trees or their products that applies to naturally growing trees in bush land. In some areas a man has full rights to the *Parkia* growing on his land, while in others the chief, or his agent, will harvest the fruit and give the farmer a proportion. Sheanut is a common resource for all community members in bush land. Disputes often arise concerning the rights over trees in fallow land.

71. Norton, A., 1988. *Participatory Forest Management in Ghana*, ODA. m.s. **GHANA** The article deals with southern Ghana and with sedentary bush swidden-fallow cultivators. Since the Local Government Ordinance of 1951, most southern Ghanaian chiefs hold the allodial title to land on behalf of the people. Hence they hold rights to all 'natural' land assets which are not the product of identifiable human labour, eg mineral rights and trees (this idea about the 'naturalness' of trees stems from the colonial period). This assignation of timber rights was an attempt to create a land-based revenue system to support the chiefly bureaucracy. This separation of tree and land tenure therefore militates, in the present day, against the successful combination of trees and crops on land. For chiefs make over their rights to timber contractors who, in exchange for royalties, are then entitled to fell any tree on farmers' properties. The timber merchants are exempt from paying damages for any crop injured in the felling, hence the farmers perceive the presence of trees among their crop as a threat. Land and tree tenure systems must be reintegrated if agroforestry projects are to be successful.

72. Onochie, C.F.A. 1964 'An experiment in controlled burning in the Sudan zone.' *Proceedings, First Nigerian Forestry Conference, Kaduna*. pps. 131-155. **NIGERIA** Experiments were conducted in *Anogeissus-Combretum* forest, (600-900mm of rainfall) on a variety of combinations of protection and burning regimes, to discover which gave the greatest Mean Annual Increment, and which the greatest number of large stems per hectare. In order of success, the greatest MAI was obtained:
1. by early burning every fourth year, and by fire protection the rest of the time.
2. by early burning and fire protection every alternating year.
3. by fire protection alone.
4. and 5. by early or late burning annually.
In order to obtain a large number of large stems per hectare (50cm and over) protection, without burning was the best regime, even though some MAI was sacrificed this way.

73. Ostberg, W. *We eat trees: Tree planting and land rehabilitation in West Pokot District, Kenya: A baseline study* Swedish University of Agricultural Science, International Rural Development Centre, Working Paper no.82, Uppsala, 1988.

KENYA Study data on West Pokot (rainfall 875mm) is used here. The people are agriculturalists who depend on livestock for security against drought. Sorghum and finger millet are the original crops with maize, pulses and beans become rapidly more important. Wild plants are picked during the rainy season as vegetables. There is clan management of grazing and farming land. Both men and women emphasise the need for fodder species. Thatching grass is in shorter supply than firewood. Men have greater knowledge about trees useful for fencing, because they are responsible for animals. Tools and weapons are made from trees. Fields are dug by hoe and fallowed for four years after 3-5 years of cultivation. Trees are maintained in the fields for dry-season fodder and are not lopped by neighbours. Fodder trees are also saved close to the house to provide fodder for sick animals kept at home when the rest are elsewhere. *Balanites aegyptiaca* is considered to be the most useful. It produces fruits in the dry season and its leaves are used as vegetable relish. It can tolerate moderate lopping. There are two types of fields: thorn fenced shamba with heavily lopped trees inside, and shamba with a living fence and moderately lopped trees. (These suggest regularly fallowed, and permanently cultivated fields respectively. Ed.) *Ficus natalensis* (simwoto) is considered sacred, its longevity and huge size encouraging people to see it as a symbol of the lineage. Thus the blessing: 'have branches and leaves like the simwoto tree'. *Ficus sycamorus* is also valued and protected. A rising population and the shift towards agriculture, are weakening the indigenous clan system and its authority, both increasing risk, and also strengthening the importance of the individual household.

74. Pélissier, P., 1966. 'Les paysans du Senegal', [The peasants of Senegal] *La civilisations agraires du Cayor à la Casamance*, 1966 pp 252-273.

SENEGAL The article describes the Serer region of Senegal, an area with an eight month dry season but no water shortage because of the high groundwater table and the presence of small lakes and rivers during the rainy season. The inhabitants are sedentary herders and millet cultivators.

'Serer parkland' comprises bushland with *Faidherbia albida* present sometimes in pure stands and also on agricultural fields. Edible fruit from *Detarium senegalense, Tamarindus indica* and *Faidherbia gnaphalocarpa* are harvested. The fruit, sap, leaves and wood of *Borassus aethiopum* are very much in demand; it is the only palm tree which is used for wine making by the Serer. *Adansonia digitata* trees seeded around villages provide food (pulp from fruit, seeds, leaves) and possess medicinal value, while the bark is used for rope making. A few other fruit trees such as mango, papaya and *Anacardium occidentale* are also present around villages. The Serer benefit greatly from the presence of regular stands of *F. albida* throughout their land. The trees, because their leaves fall during the rainy season, provide essential fodder to the livestock during the second half of the dry season, when climatic conditions are most rigorous. This is also the period when milk from the livestock is richest and butter most nutritious. These parks of *F. albida* allow for the maintenance of livestock herds at high density during the harsh dry season and provide effectively against wind erosion. In return, the livestock increase soil fertility.

F. albida has been an integral part of sedentarisation in sudanese Africa. The 'forest parks' of *F. albida* have resulted from man's initiative through their herds of livestock which eat the tree's fruit and propagate the seeds at the end of the dry season. The Serer farmer then prunes the young trees up to man's height for their protection. Thus the homogeneity and purity of a 'forest park' of *F. albida* indicates ancient occupation of the land.

75. Pélissier, P., 1979. 'L'arbre dans les paysages agraires de l'Afrique noire', pp 37-42. [The tree in the farmed countryside of Black Africa] *Le role des arbres au Sahel*, Compte rendu du colloque tenu a Dakar (Senegal) du 5 au 10 novembre 1979.

REGIONAL The article investigates trees in the agricultural crop in sub-saharan Africa. Trees are integrated with agricultural crop in Africa. From the Senegal coast to the Red Sea, tree species such as *Ziziphus jojoba, Balanites aegyptiaca, Tamarindus indica, Ficus* and *Albizia zygia* are found in association with millet, sorghum, peanuts and manioc. These 'park types' include: 'residual parks' whereby certain tree species are left standing during the first clearing for agriculture. *Cordyla pinnata* stands in Saloum and Gambie are such an example.

In the Sudanese zone, tree species like *Butyrospermum parkii* and *Parkia biglobosa* have been protected for the fruit fat content. B. parkii was selectively conserved by people who did not raise livestock (and thus could not obtain animal fats). The presence of *Elaeis guineensis* is often linked to specific cultures, sometimes very large as in Benin or in the ancient Old River States in Nigeria. *E. guineensis* naturally occurred in forest clearings which were used initially for cultivation. These 'Elaeis parks' now occur around villages everywhere in the Ivory Coast, Nigeria and Cameroon.

Borassus aethiopum is used in wine making and for its edible fruit and is found in homogeneous stands which likely originated from selective cutting. The leaves, fruit and bark of the *Adansonia digitata* are edible.

Another indigenous forest management system is that found on Mount Mandara where *Khaya senegalensis* has spread. A refined version of the system is that of the Ouldene tribe who have developed 'forest parks' of 20 to 30 trees/ha with *K. senegalensis, Ziziphus jojoba, Tamarindus indica* and *Acacia spp.* managed as domesticated plants. Also, in Serer country, sedentary farmers practising livestock raising and cereal agriculture have created *Faidherbia albida* 'parks'. Within these 'parks', *F. albida* homogeneity, population density and the importance of the herd increase at the same rate.

76. Perlov, D.C. 1984. 'Exploiting the forest: patterns and perceptions in highland Samburu'. *In*: Barnes, C. *et al* (eds.) *Wood, energy and households: perspectives on rural Kenya*. The Beijer Institute and the Scandinavian Institute of African Studies, Sweden.
KENYA The part of the Samburu area studied here contains a hot, dry grassland-and-scrub plain supporting large numbers of animals, and cooler wetter higher land supporting fewer animals, more farming, and a mix of evergreen bushland and hardwood species. The main lowland tree species found are *Acacia nilotica, Croton dichagamus*, and *A. gerardii*. On the hills are found *Juniperus procera, Olea africana, Acokanthera friensiorum* and *Croton megalocarpus*.

This paper focuses on the Samburu of one large group ranch, straddling two ecozones. Tenure was originally on clan and lineage lines, but nowadays the Kenyan Government has divided the range into group ranches for fixed subsets of relatives. The result has been greater sedentarisation and ecological specialisation. One subgroup lives close to montane forest and has few livestock; one spends most of the time out in the plain with livestock, and one lives halfway between the two. The cultivators, (usually pastoral dropouts) grow maize, beans, potatoes, and squash, but the preferred activity is still livestock keeping and trading.

Trees are used for house and granary construction, for fuel, fodder, charcoal-making, calabash cleaning (by smoking) and for medicines. There is a strong belief that it is wicked to cut green trees for fuel - because browse is so important for livestock in the dry season. There is no ownership of individual trees, but labour creates ownership, in the sense that fuelwood chopped and piled by one woman would never be appropriated by another. Samburu men believe that so long as the economic activity of the area is predominantly livestock herding, the forest will take care of itself: they identify the causes of deforestation as sedentarisation, agriculture and the consequent shrinking of pastoral lands. Permits to cut charcoal are given by the local elders and the chief's office. They discourage non-local charcoal burners by refusing them permits and if they are local Samburu give permits only for short periods to poor men **in extremis.** The same body administers fines to inhibit the exploitation of green trees within gazetted areas and within the group ranch forest reserve. The most sedentarised groups use the most fuelwood per week, and so far are making few adaptations to the fact.

77. Persson, J, 1986. 'Trees, Plants and a Rural Community in the Southern Sudan', *Unasylva* 154, Vol. 88, pps. 32-43.
SUDAN Domeri village, Western Equatoria province, Sudan, is an area of savannah woodland with patches of thicker forest. The rivers flow only in the rainy season.

The inhabitants - the Modo - are sedentary subsistence agriculturalists. They grow crops in fields and in gardens around compounds, and also fish and hunt game.

Although this article has very detailed information about the uses of trees in Domeri, the author argues that the trees are not actually managed, short of burning off to increase the fertility of the soil for crop planting. Instead, people use trees as they need them with little thought about the future. The immediate needs of crops take priority.

78. Piot, J. 1980. 'Management and utilisation methods for ligneous forages: natural stands and artificial plantations', pps. 339-349 in Houérou, H.N. le (ed.), *Browse in Africa: the current state of knowledge*. International Livestock Centre for Africa, Addis Ababa.
REGIONAL 'The classical method is that of the Sahel herdsman who used his machete or hatchet to lop trees and shrubs at head height, preferably leaving the branches attached to the trunk, giving an umbrella like habit, wholly conducive to 'roasting' and the death of the tree if fire occurs, which is still fortunately far from being the general case' (so the removal of dead branches in many natural woodlands would actually reduce the impact of bush fires on the resource - ed.) However these activities are predominantly found only in the Saharo-Sahelian zone where true nomadism is practised and rainfall is below 200mm. In the Sahel proper the need to protect crops from livestock means that hedges of thorny branches are used and the umbrella shaped trees are less frequently seen.

79. Poschen-Eiche P. 1987. *The application of farming systems research to community forestry: A case study in the Hararge Highlands, Eastern Ethiopia*

Tropical Agriculture [1] TRIOPS Verlag Tropical Scientific Books, Langen: Germany, 1987.

ETHIOPIA The area studied lies at an altitude of 1800-2500m. with rainfall of 700-1000mm/year. The area used to be covered by highland forest, with some mountain savannah, and tree-cover predominantly composed of *Podocarpus gracilor* below 2200m and *Juniperus procera* above 2200m. Sorghum is the main crop - intercropped with maize and beans. Grain is used as food, with the leaves, thin stalks and thinnings used as livestock feed, and thick stalks and stubble used for fuel. Wheat and barley are grown at higher altitudes (above 2500m), and the most important livestock are cattle, goats, sheep and donkeys.

Amhara kings from north-central Ethiopia originally conquered and introduced the feudal system; 25-50% of the harvest was claimed by the landlord; tenancy was insecure, expulsion common. 'Decisions about the use of trees and shrubs were the landlord's privilege'. In 1975, the Public Ownership of Rural Lands was proclaimed. All rural lands were expropriated, and private ownership and landlord-tenant relationships were abolished. Land was distributed to whosoever wanted to cultivate it and peasant associations were formed to administrate the process.

80. Postma, M. 1990 *Land and tree tenure in the Wolof village M'borine,Senegal.* Doctoral report for the Section of Forest Management, Department of Forestry, Wageningen Agricultural University, Netherlands.

SENEGAL The research took place in the peanut basin of Senegal. The people cultivate peanuts and manioc for sale, and millet for subsistence. The wealthy own cattle, and the rest have goats, sheep and chickens. Land near the village is permanently cultivated, and land further out is alternately fallowed and grazed for three years, and cultivated. The introduction of animal traction, and artificial fertilisers have greatly changed original swidden-fallowing practices. There are now too many residents for a return to older style fallowing, although fertilisers are again becoming harder to obtain. The only adaptation open is out-migration.

Originally, the first clearer of land (by fire) was its owner. He in turn, might grant 'rights of the axe' to individuals who wanted to use some of the land. Such rights were strong, and heritable so long as the land was never left uncultivated for more than ten years. The Wolof monarchy followed customary land usage.

However, since 1964, all land has officially belonged to the state. The local Rural Council allotted land to the tillers they found in place at the time, and it is legally impossible to buy and sell land.

Husbands allocate land annually to their wives and dependant sons, but not necessarily the same piece each year. Wives may take minor products from their husbands' fields in the rainy season, and in the dry season dead wood and shrubs from any part of the village's fields, as they may from the common area under swidden-fallow. Larger branches may only be taken from a field tree by the owner of that field, however, regardless of the season.

Indigenous tree-management in the area follows the typical pattern of the farmed parkland described by Raison (1988) with a densely settled population felling many trees, but preserving *F. albida*, *Balanites aegyptiaca*, and *Adansonia digitata*. Nowadays, however, a Government licence must be obtained for the felling of construction timber, and such licences are never granted for the felling of planted trees, exotic trees or *Faidherbia albida* trees. Tree stocks have fallen to only about 5 trees per ha. partly as a result of this stifling of indigenous fallowing methods, and partly in response to the introduction of groundnuts as a short fallow cash crop. Unfortunately, tree-protection cannot develop into tree-planting here for two reasons. Firstly, annual land allocation practices deter individuals from planting trees. Secondly, government rules banning the felling of planted trees provide no incentive to plant.

81. Poulsen, G., 1983. 'Using Farm Trees for fuelwood', *Unasylva*, Vol. 35, N° 141, pps. 26-29.

KENYA The paper describes farming practices on the slopes of Mount Kenya, where the inhabitants are sedentary permanent cultivators of coffee, maize and cowpeas, rearing small livestock. *Grevillea robusta* is the dominant farm tree in the region, exploited primarily for its timber products in the following ways:

- when trees are severely pruned, the lopped branches are used for fuelwood;
- women and children collect fallen twigs from trees and other shrub vegetation for fuelwood
- when the tree's diameter is big enough, the farmer fells the tree completely, to sell as saw timber.

G. robusta is chosen because its deep root system does not compete with the shallow root systems of crops, and it is planted to form single lines with 2-4 metres between stems, spread across fields.

Mt. Kenya farmers have an established pruning system where the trees are heavily pruned once the canopy causes shade, excess water consumption and other features detrimental to crop survival. The pruning rotation takes place at 2-3 year intervals, and the system may continue for 30-50 years. The trees are kept until they reach a certain diameter, then sold. This variant of pollarding aims at longterm conservation of the production base as well as shorter term household provision.

82. Poulsen, G., 1985a. 'Trees on Cropland: Preserving an African Heritage', *Ceres*, N° 104, Vol. 18, pps. 24-28.
CAMEROON The Maroua region of Cameroon is discussed - a lowland plain. Annual rainfall is 800mm per annum. The inhabitants are sedentary agriculturalists, producing sorghum and cotton.

Poulsen describes a *Cassia siamea* shelterbelt project begun in the mid 60s in response to donor-perceived increased desertification and population pressure. The project was carried out on farmland with the owners' permission. The project conceived of the trees solely as windbreaks, anti soil erosion devices and as a way of creating a less desiccating micro-climate by forming compact obstacles to air movement. This is why shelterbelts were selected, rather than a scatter mimicking indigenous practice.

A 30 metre 'wasteland' on either side of each belt of trees, where no crops grow, has been the result. The density of trees creates a hot airless zone within which plants perish as no wind-cooling of the ground takes place. It was observed however, that crops could grow under single trees or clusters of trees. Thus the project was forced to recognise the rationality of indigenous forms of forest management/agroforestry and indigenous spatial arrangements.

83. Poulsen, G., 1985b. 'Halting the Desert by Means of Forestry; Does it make sense?', pps. 98-108 in *Land Management and Survival*, A. Hjort (ed). Uppsala: Scandinavian Institute of African Studies.
REGIONAL The article focuses on the Sahelian zone with 200-600mm annual rainfall, and inhabitants who both herd and cultivate. Trees are used for fuelwood, small timber, gum, fruit and browse for livestock, particularly during the crop season. Important species include *Balanites aegyptiaca, Faidherbia albida, Acacia senegal,* and *Butyrospermum parkii*. In the past, Sahelians operated a classical bush swidden-fallow system, with land cropped for approximately 5 years, then left fallow for 16-20 years. Trees regenerated naturally on swidden-fallow land. There is no information given on resource management at the individual/ household level.

Poulsen argues that fallowing worked well 30 years ago, but has broken down as over-cultivation and over-grazing have taken place in response to population increases, and fuelwood and fodder shortages. He implies that there is now no resource management, as short-term considerations have taken over from long-term ones.

84. Raison, J.P. 1988. *Les 'Parcs' en Afrique: état des connaissances et perspectives de recherches* [African 'parklands' : the state of knowledge and perspectives from research] Document de Travail, Centre d'études africaines, EHESS, Paris.

REGIONAL This report investigates the parklands of Africa's Sahelian and Sudanian zones, where the inhabitants practise both agriculture and livestock raising, the latter diminishing where there are labour shortages. In these zones, settlements use a variety of constantly evolving 'parklands' placed in concentric rings around the village. There is a pattern from centre to periphery, from culture to nature, as follows:

- the village and its homegardens; here one finds a planted or transplanted parkland of *Adansonia digitata* and *Tamarindus indica;*
- the zone of permanent fields with *Faidherbia albida;*
- the zone of fields cleared in the bush, and of long swidden-fallows: *Butyrospermum parkii (Vitellaria paradoxa or karité), Parkia biglobosa* (néré) and *Ficus platyphylla.*

In the concentric ring model described for the Bwa Bobo Oulé, Upper Volta by Savonnet (1959), for instance, the village site ring is the *ka*, the *F. albida* ring is the *wa* and the *ma* is the bush swidden-fallow.

Trees are used in these areas for small timber, firewood, fodder, medicines, relishes and dietary supplements including oils, alcohol (from palms) and famine foods. Income-generating opportunities include the trading of shea butter (*Butyrospermum parkii*) gum arabic, (*Acacia senegal*) and charcoal. Small timber is important not only for building, but also for furniture making and the carving of domestic utensils. Forage is vital. Leaves are far more nutritious than grass in most of the sudanian and sahelian belt and the most important fodder trees are: *Acacia tortilis, A.senegal, A. seyal, Balanites aegyptiaca, Bauhinia rufescens, Boscia augustifolia, Maerua crassifolia.*

Although there is no mention of whether, or how, these systems are defended against encroachment by other villages, management of each type of parkland by villagers is exceptionally clearly described in this document, based as it is on another, highly specialised literature search.

a) **Parklands without a dominant tree species**

These formations indicate a densely settled population who have had to sacrifice specialised parkland because of other constraints. For instance, the Mafa of Mandara, Cameroon, who live on terraced hillsides, have to avoid shading crops on their narrow terraces so have had to abandon all the large oil-bearing species such as *Butyrospermum parkii, Diospiros mespiliformis,* and *Marcinia afzelii.* On the other hand, *Ficus spp.* are common, especially *F. gnaphalocarpa,* and *F. dicranostyla,* as leafy vegetables, and *Faidherbia albida* is common, too. Trees have taken on special roles here as sources of small timber for the carving and construction of household items, and as devices to help stop soil erosion. Raison notes a shift from management of the stand to management of the tree itself, too. Trees are pollarded, and side branches are removed in order to arrive at a sophisticated compromise between the needs for wood, soil conservation, forage and avoiding the shading of annual crops. Far too little knowledge of this single tree management was found in the wide range of references he consulted.

b) ***Ficus* parkland** Raison cites, as an example, the *Ficus*-dominated parkland of Moussey, on the left bank of Logone river in Chad, where the population depends for dietary supplements on six *Ficus spp.*, and especially *F. gnaphalocarpa*, using the fruit, and the leaves for a protein-rich vegetable relish. Fodder requirements can also be met from these *Ficus spp.* parklands.

c) ***Butyrospermum parkii* parkland - bush swidden-fallow (long fallow)** This formation is found throughout the sudanian zone apart from Senegal, and is associated with long swidden-fallows. In the untouched wooded savannah, with initially perhaps 1000 trees per ha (diameter range 5-60cm) all but 100 or so are burned so that they die. *B. parkii* are the main species preserved. Over the next four years, much of the dead wood is used up for fuelwood, as the area is cultivated, and then the whole patch is rested for 20 years and the farmer moves on outwards from his village. If one area is re-used by one village for 2-3 successive cycles, i.e.40-60 years, with the *B. parkii* protected all that time, the shade after sixty years will be so great from them all that agriculture will suffer. So the whole village will move, using the earlier area as a fruit-gathering area, and the process will start again. After 2-3 moves - 150-200 years - the *B. parkii* will be very elderly, and it will be time to start the whole process of major land clearance again, preserving younger trees.

d)***Faidherbia albida* parkland - savannah fallow (short fallow)**[2] This is the most interesting and developed of the types of parkland described. The tree, because it is in leaf in the dry season, and leafless in the rains, permits permanent cultivation under it, and also, at a density of 10-30/ha, fertilisers up to 50% of the area. Crop rotation is practised under the trees. However, it is clear that *A. albida* alone cannot restore soil fertility - animal dung is needed too. A hectare of *A. albida* parkland with twenty trees on it supports six cattle in the dry season -

---

2. Hans Ruthenberg, (*Farming Systems in the Tropics*, Third Edition, 1980, Clarendon Press, Oxford) would call the *Faidherbia albida* parklands, with their heavy dependence for proper functioning on the systematic incorporation of cattle and their dung into the system, unregulated ley systems.

enough keep the area fertile; but in the rainy season the tree is leafless, there are no agricultural residues available for fodder, and the bushland is then an essential complement. If it is nearby, livestock can browse there and be tethered on currently exploited fields at night to deposit their manure. If there is no bush nearby, more labour-intensive solutions must be adopted, such as sending animals north for part of the year, or growing forage for them. But more recently, with the departure of the young as labour migrants, animal herding has become too labour-intensive an activity. The result is often that *F. albida* parkland reverts to *Butyrospermum parkii* parkland.

The dynamics of parklands and their situation today

The construction of parkland is often the work of centuries and the clustering of trees today can be informative about management practices of the past. Inside the concentric rings of present-day parkland, one sometimes finds dense, asymmetrically placed clusters of trees, selected on completely different principles. Sometimes these hint at previous settlements, and sometimes at sacred groves containing clusters of *Balanites aegyptiaca, Ziziphus mauritianus*, and *Celtis* integrifolia. The Bissa of Burkina for instance would, until the colonial period, practise agriculture in a fine *F. albida* parkland around large, densely settled villages. Then, because of violent conquest and the introduction of forced labour, they dispersed to much smaller settlements in the bush.

However, areas in which concentric cultivation patterns are noted do seem to be fragile once population densities rise. If they cannot be abandoned because of overall population densities, they degrade. In any case, the trees have lost some of their importance. Oil-bearing trees are being replaced in the diet by groundnut, sesame and cotton-seed oil, and palm oil is far easier to extract than shea butter. There also seems to be less effort to plant famine trees such as the *Ficus spp.* and *Borassus aethiopum* though it is not clear why. The tendency is for the fallowing system to shrink to the point where it is replaced by a crop rotation of alternating millet and peanuts. Present-day land registration arrangements naturally encourage individuals to try to earmark permanent plots for themselves of the largest possible extent.

There is an excellent bibliography of published and semi-published articles on francophone African 'farmed parkland' areas. The author also writes about Guinean 'oil-palm parkland', but the section has not been included here.

85. Riesman, P. 1984. 'The Fulani in a development context: the relevance of cultural traditions for coping with change and crisis'. *Life before the drought*, E.P. Scott, ed. Allen and Unwin, Boston, pp.171-191.

NIGER The article describes the resource management practices of the nomadic and semi-sedentary Fulani, living along left bank of the Niger river. The Fulani would police the bush to prevent unauthorised cutting of branches to feed goats. They would also put out bush fires and, once the fire was under control, punish the individual(s) who had started it by a fine : the provision of a feast for the men who had put out the fire and identified the perpetrator (s).

86. Roberts. A. et.al. n.d. *The Malete law of family relations, land and succession to property* Gaborone, Government Printer.
BOTSWANA Trees situated on land which has been allotted for residential or agricultural purposes belong to the holder. Others must ask his permission before taking firewood or fruit from trees on his allocated land. He may cut them down or harvest their fruits except in the following cases: no shade tree in a residential area may be cut down except with the Chief's permission; no living tree may be cut for firewood; from the start of the ploughing season to the end of harvest, some species may not be cut at all, and others may only be cut at certain limited times. Disobedience, it was believed, would cause hail to fall on the crops.

87. Salem, B. ben, and van Nao, T. 1981. 'Fuelwood Production in Traditional Farming Systems'. *Unasylva*, Vol. 33, N° 131, pps. 13-18.
SUDAN and SENEGAL The paper discusses two systems. The Sudan case focuses on the provinces of Kordofan, Darfur, the Blue Nile and Kassala. The inhabitants with whom the study is concerned are subsistence farmers practising bush swidden-fallow. Here, *Acacia senegal* and *Balanites aegyptiaca* constitute the main tree cover, with *Acacia senegal* tapped for gum, and also used for fodder and fuelwood. In this system, crops are grown and gum produced in rotation. The stages of the cycle are as follows:

- *Acacia senegal* is cleared, *Balanites aegyptiaca* is left, and the land is cropped with *Pennisetum typhoideum* and *Sorghum vulgare* (cereals) for 4-10 years.
- *Acacia senegal* is planted or regenerated and left for 3 years.
- *Acacia senegal* dominates and is tapped for gum for 6-10 years. When the trees are aging, they are cut down and the cycle begins again.

The second example is the Western Senegal semi-arid savanna. Here annual rainfall is between 300-900mm, and the population density is between 80-100 per km^2. Tree cover is mainly *Faidherbia albida* and *Prosopis spp.* Trees are used to maintain soil fertility and moisture, the limiting factors in crop production. Trees also used for fodder and fuelwood. *Faidherbia albida* is used to improve millet production, 100 trees per ha maintaining indefinitely the croplands's nutrient requirements. There is no information on rules of management.

88. Schapera, I, 1943. *Native Land in the Bechuana Protectorate*, Cape Town, Lovedale Press.
BOTSWANA Trees and bushes are common property within the tribal territory, used for building, fencing, fuelwood, and fruit, except where another man is already cultivating. Fruiting trees and bushes in the wild may be used by anyone, but cultivated fruit trees in a homestead may not. Trees in a village, even those inside a man's homestead, may not be cut without the permission of the Chief. These trees are for shade, and to provide lavatory shelter for the old and sick who cannot go into the hills.

Certain tree species (dubbed 'female trees') might not be cut during the agricultural growing season, the taboo applying even to the uprooting of tree stumps at this time: it was feared that the rain might turn to hail and damage the crops if the taboos were not obeyed. The chief would proclaim 'closed' and 'open' seasons, and would reopen the cutting season himself, at the same time insisting that the first branches cut by ordinary people be used to repair a community institution - the cattle pound attached to the *kgotla* (the village meeting ground) - before any cutting for individual needs took place.

89. Seif el Din, A.G., 1980. 'Agroforestry Practices in the Dry Regions' in *Proceedings of the Kenya National Seminar on Agroforestry, ICRAF.*
NIGER and SUDAN Arid lands have prolonged dry seasons of 8-10 months, low and unreliable summer rainfall of 200-600mm annually, and high daytime temperatures often exceeding 40°C. Evaporation may be 20mm per day during the hottest months. Drought years occur frequently, but irregularly. The people of these regions practise livestock husbandry and crop cultivation carried out by swidden-fallowing. Two examples of these systems are given:

- The Sudanese gum arabic rotation system (described in summaries 45, 46, and 87 and not repeated here);
- The Bougagé strip rotation system, Niger.

Under this system, the household controls 17 strips of which, at any one time, 12 are being fallowed, 4 are under cultivation and one is used as a house and livestock enclosure site. Use is rotated through the 16 potential cultivation strips. Among the Bougagé, ownership rests with the household unit. No mention is made of common property resource usage.

Useful trees include *Adansonia digitata* for food, rope fibre, fruit and water storage in the hollow trunks; *Borassus aethiopum* for baskets, thatch and timber; *Faidherbia albida* for fodder; *Parkia biglobosa* for edible fruit

& shade; *Parinari macrophylla, Sclerocarya birrea* and *Anacardium occidentale* for fruit and nuts; *Balanites aegyptiaca* for fruit, fuel, shade, tools and utensils.

The above trees, and other species, are also important in their capacity to regenerate soil quality for crop production.

Each farmer bases his choice of crops to be interplanted on his prognosis of what is likely to happen climatically in the next few weeks. The elders keep an unwritten calendar divided into 28 periods, each lasting approximately 13 days and closely related to atmospheric and astronomic conditions. Land management techniques are incorporated into this system. Seif el Din argues that ecological degradation is a result of factors outside farmers' control - including population growth - rather than some flaw in their system of resource management.

90. Seif el Din, A.G., 1987. 'Gum Hashab and Land Tenure in Western Sudan', pps. 217-224 in *Land, Trees and Tenure*, J.B. Raintree (ed), Nairobi, ICRAF.

SUDAN The writer addresses the Western Sudan, Kordofan Province, an area of sandy soils and plentiful *Acacia senegal*, with cultivators growing groundnuts, sesame, hibiscus, melon seed, dura, and grazing cattle and sheep.

He argues that land tenure became an issue when the gum trade began to flourish in the late 19th Century. Tenure rules were laid down by tribal leaders, whereby group members could own land for cultivation and gum collection, as well as collect gum from the communal lands. From 1899 onwards the government has intervened in these land tenure rules, laying claim to ultimate land ownership and classing tribal cultivators as long-term tenants and lease-holders. The local land-tenure system, which villagers regard as the legitimate one, maintains that the individual has the right to gum from trees growing on currently fallow land s/he has already cultivated. But during the early part of this century the government repeatedly described the system as one of communal rather than individual tenure, where individuals had usufruct rights only.

More recently, tribal lands have been divided into village holdings in order that:

- cultivation and gum collection is restricted to people domiciled in a particular village;
- land is therefore communally-owned and inalienable;
- excess land may be allocated by head to these in need;

However, the human and animal population explosion has disrupted the bush swidden-fallow system through over-cultivation and over-grazing, and has challenged existing land tenure institutions.

People have been forced, partly by climatic changes, to move to the central gum belt as the northern areas have become unworkable.

His solution is to suggest that land tenure rules should be changed so that the government can enforce legislation protecting the environment in the long-term. Under the present communal ownership, he argues, individuals exploit the land without regard to its long-term well being.

91. Sene, E.H.M, 1979. 'L'organisation foncière dans la zone aride du Senegal et la gestion des ressources forestières', pp 15-18. [Land classification in the arid zone of Senegal, and the management of forest resources] *Le role des arbres au Sahel*, Compte rendu du colloque tenu a Dakar (Senegal) du 5 au 10 novembre 1979.
SENEGAL The establishment of a forest reserve classification during colonial times had an important impact on the local population, who were most antagonistic towards this new classification system which restricted their right to utilise soil and forest resources. In some areas trees located on farmland such as *F. albida* and *Cordyla pinnata* become a common property resource in the dry season, and may be freely pruned for fodder and utilised by herders. On the other hand, in the Casamance or Cayor areas *Borassus aethiopum* trees belong permanently to the farmer who cultivates the land.

92. Sene, E.H.M. 1985. *La participation des populations au développement forestier au Senegal*. [Popular participation in forest development in Senegal] report for FAO, Dakar, Senegal January 1985.
SENEGAL The author gives a series of examples of the vital importance of trees in Senegal. Tree species such as *Detarium, Parinari, Cordyla africana, Cola cordifolia* are protected for their edible fruit, *Bauhinia spp, Sterculia setigera* for gum and fibre production. Leaves from *Combretum micranthum* are used in the making of the quinquelibas drink and other *Combretaceae* for fuelwood production. Young *Borassus aethiopum* trees (between 15-20 yrs) are used for wine making.

Agriculture in the Serer region is based on *Faidherbia albida* and its multiple uses, including soil fertilisation and fodder. But other species have also been promoted on private land: the Serer farmer walks around his field seeding *B. aethiopum* nuts to mark important occasions such as at the birth of a child, and there have been examples of *B. aethiopum* being deliberately propagated by farmers in other parts of the country as well; farmers have maintained *Elaeis guineensis* in the Casamance area; *Adansonia digitata* groves are likely to have originated from human action; and villagers have practised the propagation by cuttings of certain species such as *Erythrina senegalensis, Lannea acida, Ficus spp.*, and

Lonchocarpus spp.

The indigenous forest management practices of local populations are decreasing as forestry has come to mean nothing but forest reserves and village forestry schemes; and while regular harvesting of palms and other species for multiple uses is practised, little maintenance is now taking place apart from nut seeding. Ironically, local people were originally the instigators of the interrelated patterns of forest and agriculture in the countryside, but, except for parts of the extreme east and south of the country, these agroforestry practices have been lost as government has become more involved.

93. Shepherd, G. 1986. 'Forest Policies, Forest Politics', *ODI Social Forestry Network Papers*, 3a.

GENERAL Customary tenure is flexible, administered by local leaders regarded as legitimate, and with a certain degree of discretion. Nationally imposed tenure systems are homogeneous throughout the country regardless of rainfall and land-use system, are fixed at one moment in time and adapt very slowly to longer-term change. A local relatively sensitive political authority is exchanged for remote, locally ignorant government.

Customary tenure states that :

- those who live nearest to natural resources have stronger rights to it than outsiders.
- self-sown trees are common property and rules governing their use are evolved locally.
- planted or preserved trees belong to the individual who has gone to the trouble to plant or protect them, and are likely to strengthen that individual's rights to the land on which the trees are found.
- the clearing of trees in heavily forested areas (i.e. the investment of labour in the creation of agricultural land) will confer tenure rights.

Modern state tenure has been able to recognise the acts of land-clearance, and planting, as acts which fit with European ideas of land title. But tree-usufruct and tree-preservation activities practised communally have been far more invisible to them, and it has followed that those who use an existing forest resource, rather than planting it or clearing it, will have weak rights. Thus the weak position today of forest dwellers. The State also, not surprisingly, finds it difficult to concede that the strongest forest rights belong to those who live nearest.

The paper highlights the present-day competition over land use which is inherent in forest problems. Conflicts over the role of trees - and hence their management - occur between local subsistence needs and State revenue demand; between rural and urban demand; and between the poor and the rich.

Once forests are under pressure, what roles must they satisfy? Dwindling forests mean that not all the functions the forest previously served can continue. What changes must be made and what can continue? Management strategies are likely to need radical revision, but the more powerful party - the State - is likely to try to maintain State functions for the forest and cancel local use-rights, though this can never work.

One solution is to allow local people to strengthen their rights in a way understood by modern state tenure systems - by planting, by land-leases, and by increasing political rights over the area they administer.

94. Shepherd, G., 1989a. 'The reality of the Commons: answering Hardin from Somalia' in *Development Policy Review*, London: Sage, Vol. 7, N° 1, March 1989, p 51-63. (The summary combines information from this paper, and from N.M. Bird, and G. Shepherd, 1989, this volume.)

SOMALIA The Bay region of Somalia lies about 300km inland from Mogadishu the capital. The area is semi-arid, receiving 300-500mm rainfall p.a. The people of the region are agro-pastoralists, herding camels, cattle and goats, and producing sorghum. Trees are vital: it is impossible to live in the area without animals, and tree-browse sustains goats and camels entirely, and cattle for much of the year. Trees also supply housing materials, fuel, agricultural and herding equipment, domestic utensils, human and animal medicines, and furniture.

Until 1960 clan membership (which gave both kinship and territorial identity) defined land rights and dictated the duty to defend lineage (sub-clan) land against outsiders. Private plots existed alongside common grazing lands, communal annually repaired dry-season rainwater reservoirs, and remoter open access areas. In 1960 clan land-rights were abolished, and both common property resources and open access land were lumped together as State land. Thereafter, nearly all previous attempts at management by area and by membership of a fixed group came to an end.

Currently, Bay groups still request permission to use each others' grazing areas as if the earlier system was in place, but have no right to ban outsiders, such as charcoal burners, from access to 'their' resources. Much of their outrage over the felling of trees for charcoal is because their own management priorities are for living trees. Senior adults discouraged excessive tree-lopping for goat fodder, and for sorghum-pit linings, and taught herd-boys to spread risk and offtake by using a wide variety of tree species for animal fodder, so that a sequence could be used right round the season; different categories of animals were taken different distances from the home base so that (for instance) species eaten by cattle unable to walk far would not be used up by camels which can travel to far more distant sites.

The proposals from villagers for their own renewed involvement in bushland management in their area were as follows:

Firstly, that a tract of bushland on the edge of the farmland be set aside for each village, from which they would have exclusive access to poles for house-building. Village chiefs or committees would arrange the protection of the reserve, who might cut poles and when, and open and closed grazing seasons.

Secondly, that village grazing areas (beyond the exclusive pole areas) be used by villagers, and by neighbouring villages on a reciprocal basis. The reciprocity is a necessary mechanism because of the highly local nature of the showers that fall from season to season and year to year.

95. Shepherd, G. 1989b. *An evaluation of the village afforestation project, Mwanza, Western Tanzania*. ODI for IIZ Austria.

TANZANIA The Sukuma live in a rolling landscape in which originally they lived in scattered homesteads around low stone outcrops, and grazed their animals seasonally in the valleys between these hills, or among the trees on their crests. Rainfall is about 750mm per annum. They are agro-pastoralists who are having to place an increasing reliance on agriculture as population densities rise. Originally, land-rights came through tribal membership and through residence. Fields were privately owned in the growing season, and reverted to CPRs unless trees were planted there, in which case they remained private property.

Sukuma rules for land-use designated each hill - and the tree-cover on it - as the exclusive common property of those who lived around it, while specific valleys were similarly used by an identifiable and fixed number of local households for their cattle. Even today, some group rules for resource use have survived, beyond and outside the rather different communal ideals of the State. For instance, in some areas there are organised bans on cutting valley grass for thatch until it has all reached a certain height. Local leaders set a date upon which everyone from the local area may go in and cut.

During the Ujamaa period, the Sukuma had to leave their dispersed homesteads and cluster in villages. This created unforeseen problems, of which one was more concentrated fuelwood gathering in limited areas, and another was lost control and management of now remote hilltop, common land and in-field tree resources. Now unprotected, these areas fell prey to urban charcoal burners.

96. Shepherd, G. et. al. 1985. for Energy Resources Limited and International Forest Science Consultancy. *A study of energy utilisation and requirements in the rural sector of Botswana*. Consultancy report prepared for the

Overseas Development Administration, U.K. and the Ministry of Mineral Resources and Water Affairs, Botswana. Vol 1 Report, vol. 2 Appendices.

BOTSWANA The study took place in eastern Botswana, an area with a semi-arid to sub-humid climate. The Tswana grow maize and beans and practise livestock rearing. The village, its chief, and the village parliament, the *kgotla*, are important institutions. Trees are used primarily for fuelwood and for house and cattle kraal const-ruction. The *kgotlas* and household compounds are also constructed of tree-trunks set upright, in imitation of cattle kraals.

Officially all land which is not privately and permanently farmed belongs to the State. However, it has been customary for particular sub-tribes within Botswana to regard local resources as their own CPRs, and to be allowed to do so. Tswana villages are permanently sited, surrounded by a grazing zone, an outer ring of bushland and ultimately, some miles away, by farmland. A village's 'lands' can be up to 15 miles away, and major cattle rearing areas even further. It was discovered that concentric circles of fuel- wood ownership, each with differing rules, surround the village.

Village amenity area. Chiefs would ban the cutting of trees within the village because they were valued for shade and cover.

Village women's fuelwood area. Within a 2-3km radius trees were regarded as the exclusive property of the village for firewood purposes, and non-villagers were chased off. Collectors (women and children) were expected to leave the most accessible wood for the elderly; had to walk straight outwards from their homes and not use the fuel resources from the other side of the village; might not take living wood. No poles might be taken from here.

Sub-tribe men's fuelwood area. Collectors with transport (always male) had to go beyond the 'collection by foot' area, and to go even further away if they were after poles. They shared this area with other men from the same tribal subsection.

Open access areas. Commercial traders from outside the area would be encouraged to travel to still remoter areas. Heavier tools were used and live trees were at times felled. However, these outsiders have little respect for indigenous zoning rules.

As the fuelwood shortage has become more acute women have tended to be freed from the task of collecting wood because it is too far away. Boys take over the job, or if a husband has transport he will collect the wood, or villagers will buy from other villagers with transport. Commercialisation is reducing the willingness to help elderly relatives or neighbours by collecting for them.

97. Sin, Abu. M.E. and El Sammani, M.O. 1987. 'Community Perception and Participation, Prospects in Management of forests in Marginal Lands of Sudan. A case study of Rawashda and Wad Kabu forests, Gedaref, Eastern region, *Geojournal*, Vol. 15, N° 4. pps. 399-411.
SUDAN The paper addresses the settled areas around Rawashda and Wad Kabu forests in the northern Sudan (for which the Fuelwood Development for Energy in Sudan Project is formulating management plans). The main land-use systems are extensive mechanised farming and intensive peasant farming and nomadism. Most people practice a combination of agriculture and livestock raising, the diversity of activities being determined by the resource base. Trees are used for charcoal and fuelwood for household, commercial and industrial use; for construction timbers; and as browse for livestock. The two forests in the study area are regarded by some villagers as common property of great value, by some as government property to be used in accordance with government regulations, and by others as tribal land appropriated by the government. This affects their plans for forest maintenance.

The authors were investigating the willingness of villagers to participate in revised forest management under government supervision. People are willing to plant in unused land, and will accept any revision of land-use if it will secure more land from mechanised farming. But they are unwilling to give part of their own land for forest expansion, mistrusting their neighbours and the government's intentions, and also prevented from doing so by the small size of their holdings. Thus proposals for village woodlots were met with suspicion, and the fear that the government wishes to lay claim to villager's land through the Forest Department.

98. Skinner, J. 1988. 'Towards better woodland management in Sahelian Mali', *ODI Social Forestry Network Papers*, 6a.
MALI The paper analyses material from the inner Niger delta (rainfall 300-450mm). The flood plain is an important grazing area, increasingly under pressure recently as rainfall has tailed off. Fishermen, cattle- and goat-herders use the area, the fishermen also growing rice and other crops.

While the plain is in flood, it is divided into fixed fishing grounds used by locally resident fisher groups. When the plain dries out and grass grows, local Peul (Fulani) herders, under their three lineage heads, the Dioros, divide the area into grazing grounds first delimited in the 19th century. The three Dioros also control the allocation of parts of the area to rice-growers and to goat herders who bring their animals to graze from up to 100 miles away. The Peul use the rangelands ; goat herders lop branches from trees to feed to their stock. Visiting fishermen, but not local fishermen, who consider the practice destructive, cut tree-branches and throw them into the river in the dry season to make fish shelters where fish are likely to be caught. Fishermen do not control herders or vice-versa.

Overarching all these management practices is the fact that Mali nationalised all land in 1960, and that the forestry staff who now administer the area are unlikely to be locals. The nearest Forestry agent lives 20km away and only visits sporadically. When he comes, he observes lopped trees and, since the individual culprit can rarely be found, imposes group fines on the goat herders. Thus fining, to the goat herders, appears to be arbitrary and they cut as they choose and pay when they have to.

Proposals include creating an area committee with representatives of the fishermen, the pastoralists and the goat herders, and registering the woodland in the name of the local village. It will then be possible to register a fixed number of herds of goats (currently 20) who will have firstcomer rights to the wood and, it is hoped, helped exclude others in the future. The Dioros are the key to the successful running of the proposed Village Committee and the Foret Villageoise, and are reckoned to be the only people who could make it work.

99. Sollart, K.M., 1986. *Traditional Tree Management and Conservation Methods in the Sahel*, Wageningen: BOS Document N° 4.

REGIONAL In the Sahel, rainfall varies between 100-600mm per annum, and the ground cover comprises, among other species, *Acacia senegal, F. albida, Panicum turgidum, Cyperus conglomeratus, Echinochlea colona, Maercia crassifolia, Balanites aegyptiaca*. Soils are fairly arid in the north, but as one moves south ferruginous soils on sands, and clay-loam soils are found. Nomads in the north and settled farmers in the south are united in the middle by a transitional belt involving both stock-raising and farming. In all systems, trees are an important component of the production base. However, according to this author, farmers are more experienced tree managers than herders are.

In much of the Sahel, tree and land ownership is distinct, which leads to conflict if the landowner regards the presence of trees as an inhibiting factor in the development of his/her crop. Thus long-term investment in planting trees is often limited to those with secure title to the land involved.

Nevertheless, tree-use is vital. Shea nuts are needed for cooking oil; bark is woven and dyed; trees produce fuelwood, fruit and fibres, medicines and construction materials. Trees offer a fall-back resource in times of drought, and tree exudates (latex, resin, gum) are important. In many agro-silvo-pastoral systems, *Acacia senegal* seedlings are protected.

In several settled farming communities, planting of trees has long been practised (e.g the gum arabic tree-swidden-fallow system). Likewise, indigenous plantations of fuelwood and fruit trees, such as *Adansonia digitata* gardens, may be found in much of the Sahel. Trees are planted for shade, fencing or religious reasons (on graves and tombs). Pruning, as a systematic management method is rarely found. Where it is, the trees are usually heavily lopped in order to provide browse. Fire is also used for management purposes.

Local taboos on certain species and activities may be deeply rooted beliefs based on historical observation and an understanding of the local ecosystem. In the case of *Faidherbia albida,* lopping is prohibited or regulated and the tree is described in proverbs and sayings as the protector of farm fields and soils. However, in Burkina Faso, there are also taboos placed on the practice of planting trees, which are connected to perceived threats to the socio-economic system. Resource management strategies have deteriorated during the post-colonial period. Among nomads and cultivators, land designated as common property has become seriously depleted as individuals and/or tribes have competitively over-exploited common lands instead of applying regulatory and conservational measures.

100. Stiles, D. and A. Kassam, 1986. 'An ethno-botanical study of Gabra plant use, Marsabit district, Kenya' *Journal of the East Africa Natural History Society and National Museum* 76 (191):1-23
KENYA The article reports on work in semi-arid north-eastern Kenya, where the Gabra, a Cushitic language speaking group, live in sparse scrub savanna. Here, individual highly valued species such as *Salvadora persica*, (used for chew-sticks/ toothbrushes) may not be cut for more mundane purposes such as fuelwood. For certain species, too, there are closed and open seasons for cutting. Reference in Niamir, 1990.

101. Swift, J. and Purata, S.E., 1987. *Forestry and food security in the pastoral economies of northern tropical Africa.* Background paper presented February 1988, Bangalore, for Expert Consultation on Forestry and Food Production/Security, FAO, Rome. (The Tuareg)
MALI The Kel Adwar, a group of Tuareg in north-eastern Mali, live in the transition zone between the Sahel and the Sahara, where rainfall is unpredictable and variable. Their main activity is pastoralism, usually with a combination of camels, cattle, sheep and goats. All animals are individually owned but land and water rights are much less clear.

Each clan is historically associated with a particular wadi or group of wadis, but within each clan, pasture is a common resource. Households using part of a wadi can prevent others from cutting down trees there, although they cannot prevent others from allowing their animals to browse on the trees.

The reduction in gathering of bush produce by the Tuareg has been attributed to the emancipation of slaves who used to perform these tasks, to the drought and to the increase in population density.

102. Thomson, J.T. 1980.*Peasant perceptions of problems and possibilities for local-level management of trees in Niger and upper Volta*, paper prepared for the African Studies Association, Philadelphia, October 1980

NIGER and BURKINA FASO This article compares an area of Upper Volta with a sub-sahelian climate and rainfall of 600-650mm, with the Mirriah, Zinder area of Niger, with a sahelian climate and rainfall of 450-600mm.

In Upper Volta (present-day Burkina Faso), sorghum and millet is grown with rice and maize. Some animals are kept, and there are orchards in valley bottoms. In Niger, sorghum and millet is grown, a wide variety of animals are kept including donkeys and horses, and nomads come through the area annually.

In Upper Volta, people had scarcely ever had dealings with the Forestry Department. They saw trees on commons, such as *Tamarindus indica, Parkia biglobosa, Faidherbia albida*, or *Adansonia digitata* as communally owned, and saw planted trees or trees regenerating on privately owned land as belonging to the planter or landowner. In Niger, where the main natural vegetation was *Acacia spp.*, rights to some species had been defined by the Forest Service. Villagers saw low-value species as their own common property, and valuable species as the property of the foresters. The same went for the trees in village woodlots. Trees were important for fuel, fodder, and poles in both areas. In Niger there was a real shortage and both poles and fuel were sold in local markets.

In Upper Volta, where contact with the forestry department had been characterised by 'benign neglect,' villagers were interested in tree-planting. They felt able to make their own decisions about whether to plant communally or individually, and were capable of working out their own management strategies. In Niger, where rules about tree-ownership are already riddled with tension because of the way in which the Forestry Code is put into practice by foresters, the prospects for any tree-planting, and especially any communal tree-planting, are virtually zero. Neither they, nor the authorities, have the power to enforce mechanisms for tree-management.

103. Thomson, J.T. 1983. *Deforestation and desertification in twentieth century arid Sahelian Africa.* prepared for conference on 'The World Economy and

World Forests in the twentieth century', University of North Carolina. **REGIONAL** In the past, group control allowed fallowing without land being lost. Present-day tenure rules, on the other hand, tend to allot land to the tiller and to re-allocate untilled land elsewhere. The result is that people become wary of fallowing their land for fear of losing it, and tree cover is lost. All trees, even those on farms, now belong on paper to the State Forestry Service rather than to the owners of fallows. Increasingly they are claimed by the farmers on whose land they are to be found, but it is now their fuel, rather than their soil amelioration qualities which are valued.

Labour migration from the Sahel has encouraged extensification of agricultural techniques, for despite population growth, more men and more young people have left the area for ever. Farmers are now trying to cultivate the most land they can in the least time, at the very beginning of the rainy season. The orderly management of land practised by the extended family in the past is collapsing, long duration swidden-fallowing is coming to an end and more and more marginal lands are farmed, with accompanying destruction of bush areas. Each household head now tries to spread his bets by sowing over as wide and varied an area as possible with the result that older practices such as manuring, intensive sowing and weeding, planned fallowing and water conservation, have all been replaced by quick easy farming. Soil fertility is declining, while health measures continue to push up population growth.

104. Thomson, J.T. 1985.*Local environmental management practices and orientations for rural forestry in Mali's fifth region.* Prepared for USAID, Bamako.

MALI Foresters were trained from 1935 to implement the forestry code and nothing else. This study went looking for indigenous management practices in Mali's fifth region and studied fourteen communities - all engaged in varying combinations in farming and herding. All applied management and conservation practices of various kinds. However, in the majority of cases promising results could not be expected from the villages managing certain of their own natural resources, without further work on their institutional capability. The most interesting for the current study were the following:

The Dogon of the Seno area used to have an elaborate system for resource management. Each village quarter designated an elder to take charge of environmental management issues. The group of elders were feared and respected and were able to impose a series of graded fines, on either co-villagers or on outsiders. The fines imposed took the form of animals to be slaughtered, the meat being distributed to all villagers.

The system is now in decline because the Forest Service has taken over patrolling and fining.

The Dogon of the Kassa area are situated in a key transhumant area. The several villages in the area mount guards to stop transhumants from hacking down branches from the adjacent bush on their way through. For most of the year the patrols have little to do, but during the northward and southward migration period, guards were mounted even at night.

105. Thomson, J.T., D.H. Feeny and R.J. Oakerson, 1986. 'Institutional Dynamics: the evolution and dissolution of Common Property Resource Management', pps. 391-424 in *Common Property Resource Management* ed. BOSTID National Academy Press, USA.

NIGER Only the case of Zinder, Niger, in the arid West African Sahel is used from this article, an area of shifting cultivation.

From 1884-1935 there were local CPRs near villages, more general CPRs further away and a nationwide commons for the tree *Faidherbia albida*. From 1935-1974 legal state protection was extended to 14 additional species, and to some extent enforced. From 1974 to 1984, CPRs failed, village woodlots as communal assets failed to replace them satisfactorily, and farmers began to claim private rights to the trees found on their farm-land.

Tree management intensified as the population increased. To start with, only a few tree species were treated as private property, with *Faidherbia albida* the preserve of the sultans. All other species were available to all. After 1935, with the colonial government's Forestry Code established, local management was curtailed and national-level rules substituted for them. The main other external changes which have had an impact on the situation have been population growth and enlarging markets.

106. Toit, R.F. du, B.M. Campbell, R.A Haney, D.Dore, 1984. *Wood usage and tree-planting in Zimbabwe's communal lands. A baseline survey of knowledge, attitudes and practices.* Report produced for The Forestry Commission of Zimbabwe and the World Bank.

ZIMBABWE A study of the miombo woodlands on Zimbabwe's communal lands, where livelihoods are based upon maize cultivation and some cattle-keeping. In deforested areas, hill-tops and riverine valley-bottoms are now the last sites from which wood for fuel, space-heating and poles can be gathered.

Communal Lands (ex-colonial native reserves) are lands outside the high quality ex-white areas where large farms exist. They are over-densely settled, and whatever management practices for woodlands in the areas may have once obtained, these have long ago died out, as the trees have disappeared under an open access regime. Formally, such lands are 'owned' by various tribal land authorities, but there is nowadays *de facto* acceptance that arable land is heritable. The majority of households have planted trees for fruit, and some have planted pole trees. Since shortages of poles are now strongly felt there have begun to be strong incentives to plant privately.

107. Vink A.T. ed. 1986. *Proceedings of the Consultative seminar on integrated forest management, with special reference to Rawashda and Wad Kabo forest reserves, Eastern Region, Sudan.* A paper from the project, Fuelwood development for energy in the Sudan. GOS, Govt of Netherlands, FAO. **SUDAN** The paper considers the situation in the natural low-rainfall woodland savannah of the Eastern region of the Sudan. The sedentary inhabitants practise agriculture and livestock rearing, and some have done well out of mechanised agriculture. Nomads migrate with their herds through the forest because other routes have been blocked by extensive tracts of grain crops. Wealthy merchants from the nearby town of Gedaref use Rawashda forest 'as a dairy farm' for their investment herds. Up till now, the forest has been a Central reserve, 'managed' for the firewood and charcoal needs of wood deficit areas much further north such as Khartoum and Omdurman. In reality management has consisted of little more than paper protection, and illegal use. Mechanised farming has caused havoc in the area; farmers clear forest with tractors and farm it for a year or two under 'taungya' permits, and then the forestry department are supposed to reseed it and protect it. In fact, however, because of decreasing rainfall, more forest than ever before needs to be reserved if the amount of fuelwood originally planned is to be produced sustainably.

The seminar with local people identified the current ways in which they used the forest (for fuel, poles, fencing, browse, and the collection of wild fruits; for dry season cash from charcoal making and for the higher protein forage that forest browse offers locally over grass) and discussed with them a new management plan to include grazing and some taungya. The seminar participants were asked about what they most wanted from the forest, the role they most wanted the forest department to fulfil, and the commitment they themselves were prepared to make to forest protection. However, it was clear even at this stage that the project hoped to secure people's co-operation in forest management without passing any control of the land to them.

The nomadic workshop participants said that while they wanted continued passage-rights for their animals through Rawashda forest, they thought foresters should deal severely with those without permits, and with illegal charcoal makers. They stated that in general they caused no damage, though during two recent drought years there had been lopping for camel browse and a high animal density in the forest. Like the nomads, agriculturalist participants said that while they had no time to share the duty of forest protection, they wanted the forest department to keep long-distance fuel transporters out and for the forest to be for local people. In addition, they felt it was better to avoid all taungya arrangements since their experience of it to date was that it led to tribal fighting and the gradual disappearance of the forest.

108. Williams, P.J. 1986. *Social assessment of the Gituza Forestry Project: evaluating local needs, perceptions and participation*. CARE International au Rwanda. 47pp.
RWANDA Northeastern Rwanda, in an area transitional between the topography of the eastern plateau, and the savanna grasslands found in neighbouring Tanzania. Most residents of Gituza commune practice subsistence farming, the major crops being beans, peas, maize and sorghum. Bananas and coffee are also grown.

A survey showed that 95% of respondents have planted trees on their farms, having collected and sown seed themselves, obtained seedlings from other farms or let trees regenerate naturally. For some species eg. *Eucalyptus spp.*, *Cyprus sp.*, and *Grevillea robusta*, people have most often obtained them from communal nurseries. Three quarters of the survey respondents have left indigenous trees on their farms. One-third have *Erythrina abyssinica*. Other species commonly left include *Vernonia amygdalina, Acacia sieberiana, Rhizinus sp.*, and *Ficus spp.* The *Vernonia amygdalina* is used for medicine and firewood, the *Rhizinus sp.* for firewood and crop (banana) supports, the other species primarily for firewood, construction and tool use. Most species were left in the fields for utilitarian reasons - interviewees were reluctant to discuss whether any species were sacred or taboo. In the older, more densely populated sectors of the commune there are few indigenous trees remaining, these having been removed to put more land under cultivation or for firewood.

109. World Bank, 1985. *Desertification in the Sahelian and Sudanian Zones of West Africa.*
REGIONAL The Sahelian-Sudanian zones of the Sahel are focused on. There are three main land-use systems:

- Agro-silvicultural system (eg Hausa, Mossi). Here, production is based on rainfed agriculture in association with trees and shrubs, on a sedentary basis. In agro-silvicultural systems, trees are used for fuelwood; construction poles; bark for cord and medicine; thorny fencing; grasses for thatching, fodder and green manure; fruit and leaves. Trees are also used to protect the soil against wind and water erosion and regenerate soil fertility in bush swidden-fallow systems. Micro-organisms associated with root systems fix nitrogen and increase the productivity of crops.

- Agro-silvo-pastoral system (eg Senegalese Serer, Nigerien Bugagé). Here, livestock production is integrated into the crop/tree/shrub associations found in agro-silvicultural systems, in order to reinforce soil fertility. In agro-silvo-pastoral systems, trees are mainly used to produce browse or seed pods suitable for fodder; they are also exploited for gum arabic, nut butter, other food products, fuelwood and building material.
- Silvo-pastoral system (eg Tuareg). Large herds are maintained, through seasonal movement, to find forage and browse. Trees are a critical fodder source in the dry season. In silvo-pastoral systems, trees are primarily a source of fodder, particularly in the dry season. Tenure rules in these systems are as follows:
- In the agro-silvicultural system, land is allocated initially on a usufruct basis, then becomes subject to life tenure, then heritable property and finally saleable/mortgageable. Ownership rests initially in the kin-group, then the household and finally the individual as population pressure on available resources increases.
- In the agro-silvo-pastoral system the evolution of land tenure is similar to that in agro-silvicultural systems.
- In silvo-pastoral systems, land tenure relations originally reflected the relative political and military power of different herding groups, and were therefore competitive. However, within the group-owned tract of land, pasture was treated as regulated common property, to which all group herds had access.

In none of the systems has deliberate planting of trees constituted a significant part of indigenous resource management, as most areas initially enjoyed a surplus of natural forest cover. However, selective cutting and management of the regeneration of natural forest cover often produced stands of preferred tree/shrub species in the fields; and in the bush certain species were also favoured. Trees were protected as much for their 'social' and household uses (ie gum, honey, medicines) as their utility in providing browse, and timber, and recycling nutrients. So a deliberate policy of enrichment has taken place. Rules regarding proper use were enforced by kin-group heads.

Monetarisation and involvement in the wider political economy undermined the authority of the original kin-group structure, at the same time changing the emphasis on subsistence to cash cropping and the sending of remittances. These changes in kin-group cohesion and patriarchal authority were concomitant with an increasingly unregulated exploitation of land, and have gradually supplanted indigenous practices of soil, pasture, tree and water conservation. Unfortunately, these changes occurred during the abnormally wet period of 1950-1965, so that the initial effect of resource management changes were cushioned and masked by good rains. The 1968-73 drought exposed the breakdown in original kin structures and their careful resource management practices.

Increasingly centralised political authority has challenged the capability of local decision-making bodies to manage their environment; and government legislation becomes necessary for the smallest changes to established practices, and new initiatives, dissuading groups from organizing. Urban-biased economic policies are also isolated by the Bank as bearing responsibility for further extensive agricultural production and the shortening of fallow periods, because of low producer prices. Likewise, the failure to update stumpage fees in real terms since independence, in order to bring them more into line with the cost of planting trees, has promoted destruction of the natural forest cover with no corresponding investment in replanting.

110. World Resources Institute 1987. 'Elements of Success: Sustainable Development in Sub-Saharan Africa', in *World Resources*, pps. 221-238. **REGIONAL** The article deals with Sub-Saharan Africa. It notes the problems of its soils: of low fertility, derived from ancient weathered rocks, highly leached, low in nutrients, and with a low clay content. They are highly susceptible to erosion once the vegetation cover is removed. Shifting cultivation has been an important adaptation to these constraints, but it depends, for success, on abundant land.

Colonial and independent governments nationalised much natural forest, excluding people from exercising their original rights. This change in tenure encourages people to poach from and exploit it in response to exclusion. The author suggests that long-standing usufruct rights be restored in order to restore in turn indigenous strategies of resource management. Forests must become common property, to be managed for the common good.

Recent development projects in Niger and Nigeria have initiated systems of windbreaks and alley farming respectively, both of which have been successful. Both agroforestry projects have changed the indigenous resource management rationale, whereby the ecological benefits of trees for the soil accrue to crops only after the trees have been cleared. Instead, in the alley cropping system, where food crops are grown between regularly pruned hedgerows of fast-growing nitrogen-fixing trees, a variety of benefits accrue while the trees continue their growth and are seasonally pruned.

111. Wormald, T.J. 1984. *The management of the natural forests in the arid and semi-arid zones of East and Southern Africa*. A report for ODA.

REGIONAL There are two types of forest in this large area: the northern Somali-Maasai *Acacia-Commiphora* deciduous bushland, sparse except by rivers and with sparse grass cover; and the southern *Miombo* (*Brachystegia spp.*) with a vigorous grass layer. In both areas, large tracts of forest have been preserved by the presence of tsetse fly. Yields are low - perhaps 0.3cu.m./ha/yr. There is a natural cycle of degradation and renewal as a response to irregular rainfall. In the northern woodlands, animal herding is the chief economic activity, with some agriculture; in the southern woodlands, agriculture is the main activity of a more sedentary population, with some animal raising.

Unfarmed land in both areas has been held in the past under communal lineage or clan arrangements, but has often been transformed into State land in the last 3-4 decades, with a resulting reduction in local interest in sustainable management: clearcut tenure being an absolute prerequisite for good local management. Trees are vital in both areas for browse, fuelwood, building materials, fibres and soil fertility, with honey and charcoal important for income generation. In addition, the *Acacia-Commiphora* woodlands are important for a variety of gums, exudates, relishes, dietary supplements and medicines.

The woodlands have been maintained and kept open by human intervention over a long period. Grazing keeps the balance between annual and perennial grasses, and between grassland and bush, and is instrumental in reducing inopportune fires. Fire is also used as a management tool, particularly in the *miombo* areas, but harm can be caused in more fragile situations. There is now much more noncyclical degradation to be seen, resulting from excessive clearing: for cultivation as population density rises; for mechanised agricultural schemes such as those of the Sudan; and as a result of large-scale population movements of refugees in North East Africa. Animal herds are also crowded into smaller graze and browse areas as a result of large-scale clearing for agriculture.

Wormald makes certain management proposals : since plantations are respected, it might be possible in some areas to protect forest by planting plantations around it; it is important to spread offtake over as wide an area as possible, and to diversify the intended returns from the forest. Good extension systems are needed, based on understanding of the social and economic problems as well as the ecological ones and making sure that villagers have an economic stake in good management. The report contains an excellent bibliography.

Country Index

Annexes

Annex 1

Characteristics of the Sahelian and Sudanian zones

***Sahel zone* (Glantz) (or *Sahelo-Saharan*, Bonkoungou)**

Annual rainfall	**200-400mm**
General description	'steppe, thorny semi-desert scrub' (Jackson)
Typical tree-species	Below 250mm *A. tortilis*. At 250mm *A. senegal*, *A.seyal* and more ground cover. (Jackson)
Standing vol/ha	0-5-6 cu.m/ha
MAI	**acc. to Clément, 1982, below: 0.3 cu.m /ha**
Human activities	Agriculture concentrated near water, animal rearing predominant, nomadic or transhumant. Forage production takes precedence.

***Sudano-Sahelian subzone* (Glantz) (or *Sahelian*, Bonkoungou)**

Annual rainfall	**400-600mm**
General description	'shrub savannah to steppe' True 'shrub savanna' has less than 40 trees per ha, heights are under 7m and diameters only 7-10cm. (Jackson)
Typical tree-species	*Balanites aegyptiaca*, *Acacia seyal*, Boscia spp. *A. senegal*, *A. albida*, and non-thorny *combretaceae*. *Guiera senegalensis*, *Sclerocarya birrea*. (Jackson)
Standing vol/ha	3-10-12 cu.m./ha
MAI	**acc. to Clément, 1982, 0.06-0.5 cu.m./ha with a median of 0.3 cu.m./ha at 500mm rainfall.**
Human activities	Agricultural development, mainly in cooler areas, with animal rearing dominant, mainly transhumant. Wood or forage may take precedence, depending on conditions.

Sudan zone (Glantz) (or *Sudano-sahelian* or *Sudan savannah*, Bonkoungou)	
Annual rainfall	**600-800mm (Sudan savannah 510-1140mm)**
General description	'tree to shrub savannah' True 'tree savanna' has 40-42 cu.m/ha standing volume and at least 40 trees per ha over 7m high and over 10cm in diameter. (Jackson)
Typical tree-species	*Faidherbia albida, Bombax costatum, Sterculia setigera, Anogeissus schimperi, Prosopis africana, Terminalia laxiflora.* In this zone lots of cultivation, with *Butyrospermum parkii, Parkia spp.* and *Tamarindus indica.* (Jackson)
Standing vol/ha	5-18-20 cu.m./ha.
MAI	**acc. to Clément, 1982, 0.12-0.9 cu.m./ha with a median of 0.6 cu.m/ha for 700mm rainfall.**
Human activities	Intensive agriculture (millet, sorghum, peanut,) part transhumant, part stationery animal rearing. Wood more important than forage production.

Annex 2

Additional French bibliographic materials from J. P. Raison, 1988
Les parcs en Afrique: état des connaissances et perspectives de recherches
Document de Travail, Centre d'études africaines, EHESS, Paris.

Anon. 1975. *Underexploited tropical plants with promising economic value. Report of and ad hoc panel of the advisory committee on technology innovation*, Washington D.C., National Academy of Sciences, 188 p.

Anthéaume, B., 1972. 'La palmeraie du Mono: approche géographique', Paris, *Cahiers d'Etudes Africaines*, XII-47, p.458-484.

Anthéaume, B., 1978. *Agbetiko, terroir de la basse vallée du Mono Sud togo)*, Paris, ORSTOM, p. 126.

Aubreville, A., 1950. *Flore forestière soudano-guinéenne, A.E.F.*, Paris, Société d'Editions Maritimes et Coloniales, 523 p.

Barral, H., 1968. *Tiogo (Haute Volta)*, Paris, ORSTOM, 64 p.

Benoit, M., 1977 a. *Introduction à la géographie des aires pastorales soudaniennes*, Paris, ORSTOM, 94 p.

Benoit , M., 1977 b. 'Mutation agraire dans l'ouest de la Haute Volta. Le cas de Daboura (sous-préfecture de Nouna)', Paris, *Cahiers ORSTOM, Série Science Humaines*, XIV-2, p. 95-111.

Bergeret, A., 1983. *La cueillette: pratique périmée ou gage de santé? Etude dans trois ethnies du Saloum (Sénégal)*, Paris, DGRST, ERA 773, 101p.

Bienayme, A., Servant, M. et Desmarest, J., 1956. 'La conservation des amandes de karité', Paris, *Oleagineux*, octobre 1956.

Bille, J. C.. 1971. *Observation préliminaires sur quelques arbres du Sahel sénégalais*, Dakar, ORSTOM, 49 p.

Bille, J. C., 1978. *Rôles des arbres et arbustes en tant que sources de protéines dans la gestion des pâturages d'Afrique tropicale*, Jakarta, Eighth World Forest Congress, 21 p.

Bolza, E. et Keating, W.G., 1972. *African timbers: the properties, uses and characteristics of 700 species*, Melbourne, Commonwealth Scientific and Industrial Research Organ, 700 p.

Boulet, J., 1975. *Magoumaz. Pays mafa (nord Cameroun)*, Paris,

OSTROM, 92 p.

Boutrais, J., 1978. 'Peuplement et milieu naturel en zone soudanienne: le cas de la plaine Koutine', Paris, *Cahiers de l'ORSTOM, Série Sciences Humaines,* XV-2, p. 103-143.

Boutrais, J. (ed) 1984. *Le Nord du Cameroun. Des hommes. Une région,* Paris, ORSTOM, 551 p.

Boutrais, J. 1973. *La colonisation des plaines par les montagnards au nord du Cameroun (monts Mandara),* Paris, ORSTOM, 279 P.

Brasseur, G. 1953. *La palmeraie de Porto-Novo,* Dakar, IFAN, Mémoire no. 32, p. 31-151.

Busson, F. 1965. *Plantes alimentaires de l'Ouest africain, étude botanique, biologique et chimique,* Paris, Ministère de la Cooperation, Ministère d'Etat a la Recherche Scientifique et Technique, Ministère des Armées, 568 p.

Capron, J. 1973. *Communautés villageoises bwa, Mali, Haute Volta,* Paris, Muséum National d'Histoire naturelle, Institut d'Ethnologie, Mémoire no. IX.

Champaud, J. 1973. *Mom, terroir bassa,* Paris, OSTROM, 60 p.

Charreau, C. et Vidal, P., 1965. 'Influence de l'Acacia albida Del. sur le sol, la nutrition minérale et les rendements des mils Pennisetum au Sénégal', Paris, *L'Agronomie Tropicale,* XX, 6-7, p. 600-626.

Chevalier, A., 1932. 'Les jujubiers ou Zizipus de l'ancien monde et l'utilisation de leurs fruits', Paris, *Revue de Botanique Appliquée et d'Agriculture tropicale,* 12, p. 133-134.

Chevalier, A. 1949. 'Répartition géographique et exploitation des palmiers Borassus', Paris, *Revue de Botanique Appliquée et d'Agriculture tropicale,* p. 585-592.

Chevalier, A. et Perrot, E. 1911. *Les Kolatiers et la noix de kola,* Paris, Challamel.

Chevalier, A. et Dubois, R. 1938. 'Les palmiers Hyphaene et Borassus de l'Afrique occidentale', Paris, *Revue de Botanique Appliquée et d'Agriculture tropicale,* p. 93-102.

Clamagirand, B., et Bruffaerts, J.C. 1983. *Evolution ou disparition d'une technologie villageoise: le Karité,* Paris, CEPAZE, 2 t.143 p.

Clanet, J.C. et Gillet, H. 1980. *Le Commiphora africana, véritable arbre fourrager sahélien,* Addis-Abbeba, Colloque sur les fourrages ligneux en Afrique.

Creach, P. 1940. 'Le Balanites aegyptiaca, ses multiples applications au Tchad', Paris, *Revue de Botanique appliquée et d'Agriculture tropicale*, p. 578-593.

Dalziel, J. M. 1937. *The useful plants of West Africa*, Londres, Crown Agents for the Colonies, 612 p.

Dancette C. et Poulain, J. F. 1968. *Influence de l'Acacia albida sur les facteurs pédoclimatiques et les rendements des cultures. Nouvelle contribution*, Bambey, CNRA, 45 p.

Delwaulle, J. C. 1977. 'Le rôle de la foresterie dans la lutte contre la desertification et sa contribution au devéloppement', Nogent, *Bois et Forets des Tropiques*, 152, p. 55-60.

Delwaulle, J. C. 1978-1979. 'Plantations forestières en Afrique tropicale sèche. Techniques et espèces a utiliser', Nogent, *Bois et Forêts des Tropiques*, 181, p. 15-28, 182, p. 3-18, p. 3-17, 184, p. 45-59, 185, p. 3-23, 186, p.3-34, 187, p. 3-30, 188, p.3-29.

Delwaulle, J. C. et Roederer, Y. 1973. 'Le bois de feu a Niamey', Nogent, *Bois et Forêts des Tropiques*, 152, p. 55-60.

Depierre, D. 1969. 'Les expériences de gommeraie cultivée et leurs enseignements au Tchad', Nogent, *Bois et Forêts des Tropiques*, 125, p. 27-34.

Doat, J.C. 1977. 'Pouvoir calorifique des bois tropicaux', Nogent, *Bois et Forêts des Tropiques*, 172, p. 341-352.

F.A.O. 1977. *Boisement des savanes en Afriques*, Rome, FAO, 353 p.

F.A.O. 1956. *Acacia albida Del. pro syn. Faidherbia albida (Del. A. Chev.) in tree planting practices in tropical Africa*, 137 p.

Gallais, J. 1967. *Le delta intérieur du Niger. Etude de géographie régionale*, Dakar, IFAN, 2 tonnes, 626 p.

Gallais, J. 1975. *Paysans et pasteurs du Gourma. La condition sahélienne*, Bordeaux, CEGET, 239 p.

Gerling, Chr. 1982. *Quide de terrain des ligneux soudano-guinéens*, Wageningen, H. Veenman et Zonen B. V., 340 p.

Giffard, P.L. 1964. 'Les possiblités de reboisement en Acacia albida au Sénégal', Nogent, *Bois et Forêts des Tropiques*, 95, p. 29-33.

Giffard, P.L. 1966. 'Les gommiers: Acacia sénégal Willd., Acacia laeta R. Pr.' Nogent, *Bois et Forêts des Tropiques*, 105, p. 21-52.

Giffard, P.L. (1967. 'Le palmier rônier Borassus aethiopum Mart.', Nogent, *Bois et Forêts des Tropiques*, 116, p. 3-13.

Giffard, P.L. 1968. *Utilisation de Acacia albida dans la régénération des sols en zones tropicales arides*, Madrid, 6th World Forest

Congress, Proceedings, Vol. 3, p. 3195-3197.

Giffard, P.L. 1969. *Recherches effectuées au Sénégal sur Acacia albida,* Dakar, CTFT.

Giffard, P.L. 1971. 'Recherches complémentaires sur Acacia albida Del.', Nogent, *Bois et Forêts des Tropiques,* 135, p. 3-20.

Giffard, P.L. 1973. *La sylviculture des gommiers,* Dakar, CTFT, 28 p.

Giffard, P.L. 1974 a. *Les essences de reboisement au Sénégal. Le Kad, Acacia albida,* Dakar, CTFT, 34 p. multgr.

Giffard, P.L. 1974 b. *L'arbre dans le paysage sénégalais. Sylviculture en zone tropicale seche,* Dakar, CTFT, 431 p.

Giffard, P.L. 1975. 'Les gommiers, essences de reboisement pour les régions sahéliennes', Nogent, *Bois et Forêts des Tropiques,* 161, p. 3-21.

Gilg, J.P. 1970. 'Culture commerciale et discipline agraire. Dobadéné (Tchad)', Paris, *Etudes Rurales,* 37-38-39, p. 173-197.

Guillard, J., 1965. *Golonpoui, nord Cameroun,* Paris-la Haye, Mouton 498 p.

Hall, A.E., Cannell, G.H. et Lawton, H.W. (eds.), 1979. *Agriculture in semi-arid environments,* Berlin-Heidelberg-New York, Springer Verlag, 340 p.

Hervouet, J.P., 1979. 'Organisation de l'espace et épidémiologie de l'onchocercose', in *Maîtrise de l'espace agraire et développement en Afrique tropicale. Logique paysanne et rationalité technique. Actes du Colloque de Ouagadougou,* Paris, ORSTOM, p. 179-189.

Houerou, H.N., 1979. *Le rôle des arbres et des arbustes dans les pâturages sahéliens,* Addis-Abeba, CIPEA, 35 p.

Houerou, H.N., 1980. *Le rôle des ligneux fourragers dans les zones sahélienne et soudanienne,* Addis-Abeba, CIPEA, Colloque international sur les fourrages ligneux en Afrique, 50p.

Hurault, J. et Vallet, J., 1963. *Mission d'études des structures agraires dans le Sud Dahomey,*Vincennes, IGN, 74 p. and maps.

Jacques-Felix, H. 'Le Karité (Butyrospermum parkii) au Cameroun', Paris, *Revue de Botanique Appliquée et d'Agriculture tropicale,* no 279-280.

Jung, G., 1966. *Etude de l'influence de l'Acacia albida sur les processus micro-biologiques dans le sol et sur leurs variations saisonnières,* Dakar, ORSTOM.

Kane, N.O., 1971. *Note sur les problèmes de sylviculture et d'aménagement des gommiers,* Nancy, ENGREF, Rapport de stage.

Kerharo, J. et Adam, J.G., 1974. *La pharmacopée sénégalaise traditionnelle. Plantes médicinales et toxiques,* Paris, Vigot frères, 1011 p.

Kerharo, J. et Bouquet, A., 1950. *Plantes médicinales et toxiques de la Côte d'Ivoire-Haute Volta,* Paris, Vigot frères, 295 p.

Lebrun, J., 1968. 'A propos du rythme végétatif de l'Acacia albida Del', *Colletc. Bot.,* 7, p. 625-636.

Lericollais, A, 1972. *Sob.Etude géographique d'un terroir sérer,* Paris, ORSTOM, 110p.

Lovejoy, P.E., 1980. 'Kola in the history of West Africa', Paris, *Cahiers d'Etudes Africaines,* XX, 1-2, p. 97-134.

Ludwig, H.D., 1968. 'Permanent farming on Ukara. The impact of land shortage on husbandry practices', in Ruthenberg, H., ed., *Smallholder farming and smallholder development in Tanzania,* Munich, Weltforum Verlag, p. 87-135.

Marchal, J.Y., 1980. 'Arbres et brousses dans le paysage soudano-sahélien. Dynamique des formations végétales au nord de la Haute Volta', Paris, *Cahiers de l'ORSTOM, Série Sciences Humaines,* XVII, 3-4, p.137-149.

Marchal, J.Y., 1983. *Yatenga. Nord Haute Volta. La dynamique d'un espace rural soudano-sahélien,* Paris, ORSTOM, 873 p. and 1 vol. of maps.

Marzouk-Schmitz, Y., 1984. 'Instruments aratoires, systèmes de cultures et différenciation intra-ethnique', Paris, *Cahiers de l'0RSTOM, Série Sciences Humaines,* XX, 3-4, p. 399-425.

Maydell, H.J. von, 1983. *Arbres et arbustes du Sahel. Leurs caractéristiques et leurs utilisations,* Eschborn, Schriften reihe der GTZ, n°147, 531 p.

Mondjannagni, A., 1977. *Campagnes et villes au sud de la république de Bénin,* Paris, Mouton, 616 p.

Nebout, J.P. et Toutain, B., 1978. *République de Haute Volta: Etude sur les arbres fourragers dans la zone sahélienne,* Nogent-Maisons-**Alfort, CTFT-IEMVT,** 119 p. multigr.

Niang, M., 1975. 'Le rônier dans la région de Thiès. Etude géographique', Dakar, *Notes Africaines,* p.77-82.

Niang, M., 1981. *Place de trois arbres (Acacia albida, Adansonia digitata et Borassus flabellifer) dans les systèmes de production et de consommation des populations rurales de la région de Thiès* (Sénégal), Dakar, ISRA, 12 p. multigr.

Nicolai, H., 1963. *Le Kwilu. Etude géographique d'une région congolaise*, Bruxelles, CEMUBAC, 472 p.

Nicolas, P., 1977. *Contribution à l'étude phytogéographique de la forêt gabonaise*, Paris, EHESS, 351 p. multigr. and annexes.

Pelissier, P., 1954. *Types et genèse des paysages de parc élaborés par l'agriculture africaine*, Londres, XX° Congrès International de Géographie.

Pelissier, P., 1963. *Les pays du Bas Ouémé. Une région témoin du Dahomey méridional*, Dakar, Travaux du département de Géographie, 173 p.

Pelissier, P., 1963. *Les paysans du Sénégal. Les civilisations agraires du Cayor à la Casamance*, St Yrieix, Fabrègues, 974 p.

Pelissier, P., 1980. 'L'arbre dans les paysages agraires de l'Afrique noire', Paris, *Cahiers de l'ORSTOM, Série Sciences Humaines*, XVII, 3-4, p. 131-136.

Peltre-Wurtz, J., 1984. 'La charrue, le travail et l'arbre', Paris, *Cahiers de l'ORSTOM, Série Sciences Humaines*, XX, 3-4, p. 633-343.

Philip, M.S., 1974. 'Gum arabic and Acacia senegal', *Nigerian Forestry Information Bulletin*, 29, 12 p.

Porteres, R., 1964-65. 'Le palmier rônier Borassus aethiopum dans la province baoulé (Côte d'Ivoire)', Paris, *Journal d'Agriculture tropicale et de Botanique appliquée*, XI et XII.

Poulain, J.F., et Dancette, C., 1968. *Influence de l'Acacia albida sur les facteurs pédo-climatiques et les rendements des cultures*, Bambey, CNRA.

Pradeau, Chr., 1970. 'Kokolibou (Haute Volta) ou le pays dagari à travers un terroir', Paris, *Etudes Rurales*, 37-38-39, p. 85-112.

Pullan, 1974. 'Farmed parkland in West Africa', *Savanna*, III-2, p. 119-141.

Radwanski, S.A. et Wickens, G.E., 1967. 'The ecology of Acacia albida onmantle soils in Zalingei, Jebel Marra, Sudan', Oxford, *Journal of Applied Ecology*, 4, p 569-579.

Ruyssen, B., 1957. 'Le Karité au Soudan', Nogent, *L'Agronomie tropicale*, XII, 2-3-4.

Sahni, K.C., 1968. *Impartant trees of the northern Sudan*, Khartoum, PNUD-FAO, Forest Research and Education Centre, 138 p.

Sautter, G., 1962. 'A propos de quelques terroirs d'Afrique occidentale. Essai comparatif', Paris, *Etudes Rurales*, 4, p.24-86.

Sautter, G., 1968. *Les structures agraires en Afrique tropicale,* Paris, CDU, 267 p. multigr.

Savonnet, G., 1959. 'Un système de cultures perfectionné pratiqué par lees Bwaba-bobo Oulé de la région de Houndé (Haute Volta)', Dakar, *Bulletin de l'IFAN,* B, 21, p. 425-458.

Savonnet, G., 1980. 'L'arbre, le fruit et le petit berger du Lobi', Paris, *Cahiers de l'ORSTOM, Série Sciences Humaines,* XVII, 3-4, p.227-234.

Seignobos Chr, 1982. 'Matières grasses, parcs et civilisations agraires. Tchad et Nord Cameroun', Bordeaux, *Cahiers d'Outre-Mer,* 35-139, p. 229-269.

Seignobos Chr., 1982. 'Végétations anthropiques dans la zone soudano-sahélienne: la problématique des parcs', Yaoundé, *Revue de Géographie du Cameroun,* 3-1, p. 1-23.

Servant, M., Desmarest, J. et Poliakoff, J., 1956. 'L'amélioration du Karité au stade de la production paysanne', Paris, *Oléagineux,* February issue.

Toutain, B., 1978. *Inventaire floristique du Sahel de Haute Volta et du nord du pays gourmantché. Ecologie des plantes, noms vernaculaires, intérêt fourrager,* Maisons-Alfort, IEMVT, 122 p.

Touzeau, I., 1973. *Les arbres fourragers de la zone sahélienne de l'Afrique,* Toulouse, Ecole Nationale Vétérinaire, Dissertation, 125 p. multigr.

Wickens, G.E., 1969. 'A study of Acacia albida Del. (Mimosoideae)', London, *Kew Bulletin,* 23, p. 181-202.

Wispelaere, G. de, 1980. 'Les photographies aériennes témoins de la dégradation du couvert ligneux dans un géosystème sahélien sénégalais. Influence de la proximité d'un forage', Paris, *Cahiers de l'ORSTOM, Série Sciences Humaines,* XVII, 3-4, p. 155-166.